Plant Breeding

NIPA GENX ELECTRONIC RESOURCES & SOLUTIONS P. LTD.
New Delhi-110 034

About the Author

Dr. A.K. Sharma is Professor and Head, Department of Genetics and Plant Breeding and Department of Biotechnology COA, SKRAU, Bikaner. He is also are Arid Legumes Breeder. Professor Sharma has published 75 research papers in various National and International repute journals besides participated in many National and International conferences. He has published six books, two bulletins, four folders, five practical manuals, and two research books. Professor Sharma has guided number of M.Sc. and Ph.D. students as a major advisor, actively engaged in teaching the courses of UG, PG and Ph.D. of Genetics and Plant Breeding and handles two Research Projects as PI. He has conducted three Webinar on various aspects as organizing secretary. He is member of the Academic council of the university.

Plant Breeding
Fundamentals and Applications

A.K. Sharma
Professor and Head
Department of Genetics and Plant Breeding
College of Agriculture
S.K.Rajasthan Agricultural University
Bikaner (Rajasthan)

NIPA GENX ELECTRONIC RESOURCES & SOLUTIONS P. LTD.
New Delhi-110 034

NIPA® GENX ELECTRONIC
RESOURCES & SOLUTIONS P. LTD.

101,103, Vikas Surya Plaza, CU Block
L.S.C. Market, Pitam Pura, New Delhi-110 034
Ph : +91 11 27341616, 27341717, 27341718
E-mail: newindiapublishingagency@gmail.com
www: www.nipabooks.com

For customer assistance, please contact
Phone: + 91-11-27 34 17 17
Fax: + 91-11- 27 34 16 16
E-Mail: feedbacks@nipabooks.com

ISBN: 978-81-19235-55-1

Composed and Designed by NIPA®.

Preface

Plant breeding involves the systematic production of crop populations exhibiting genetic segregation and selection within that population to establish lines with favorable allele combinations. In case of cross-pollinating crop plants this is accomplished by recurrent selection of heterozygous plants, whereas for self-pollinating crops, pure-line breeding and other breeding approaches are preferred. Breeding of vegetatively-propagating species presents special challenges and relies heavily on culture techniques and induction of somatic mutations. Wide hybridizations, tissue culture and mutagenesis are employed by breeders to generate new alleles and broaden available genetic resources. Molecular markers are used to assist breeders through marker-assisted selection and to identify quantitative trait loci (QTL) for traits of interest. Transgenic technologies are being increasingly used as a rapid, targeted means of introducing new genes, or introducing desirable novel alleles.

A number of exquisite knowledge books on plant breeding are available which provide sufficient knowledge about various aspects of plant breeding. However, purpose of this book is to compile and provide the up-to-date information about various principles and their implication in plant breeding. The contents of this book are dealt with in 20 chapters, which cover a wide range of topics. I have consulted a large number of available books, research papers, review articles and the material available at internet on plant breeding. The ultimate credit of those endeavors should go to all the esteemed authors and scientists whose valuable contribution and inputs has been included in this book.

Thus this book would serve not only undergraduate but postgraduate students of plant breeding and agricultural botany of various universities. My sincere acknowledgement to my Ph.D. students, Swarnlata, Sneha Gupta, Komal Shekhawat and Ramandeep Kaur for providing me necessary help for checking the manuscript. Lastly, I acknowledge very sincere encouragements to my wife Mrs. (Dr.) Archana Sharma.

I hope this book will be useful to the students, teachers and researchers in the field of plant breeding and also those preparing for competitive examinations.

I invite healthy and concrete suggestion to rectify any errors or suggestions for further improvement of this book. I am grateful to New India Publishing Agency, books and their production staff for bringing out this book in an excellent form.

A. K. SHARMA
S.K. Rajasthan Agricultural University
Bikaner

Contents

Glossary

A

A Line: The male-sterile parent line used to produce hybrid seed.
Abiotic: Adverse effects on host due to environmental factors.
Actinomorphic: Flower, which can be divided into similar halves along two or more plane.
Apogamy: Development of embryo from synergids or antipodal cells.
Apomixis: Development of seed without fertilization.
Adaptation: An inherited or acquired modification in an organism that makes it better suited to survive and reproduce in a particular environment.
Additive variance: Magnitude of the genetic variance that results from the additive actions of genes, the value that the genetic variance would assume if there were no dominance or epistasis.
Allele: Alternate form of a gene.
Allopolyploid: A polyploid individual which originates by combining complete chroomosome sets from two or more species.
Alpha rays: They are composed of two protons and two neutrons, and have double positive charge. They are generated by the isotopes of heavier elements.
Aneuploidy: Chromosome number is not an exact multiple of the haploid number produced by loss or duplication of chromosomes or chromosomal regions usually due to non-disjunction.
Anther: The pollen-bearing portion of the stamen.
Anthesis: Opening of the flower.
Apogamy: Development of an individual from a gametophytic cell (such as synergids or antipodal cells) other than the egg cell without fertilization. Such individuals are haploids.
Artificial selection: Selection imposed by a breeder as opposed to natural causes, in which individual organisms of only certain phenotypes are allowed to breed.
Assortative mating: Non-random selection of mating partners with respect to one or more traits; it is positive when like phenotypes mate more frequently than expected by chance.
Avirulence: The inability of a pathogen to cause a disease.
Auxin: Induces cell division, cell elongation, swelling of tissues, formation of callus, formation of adventitious roots, inhibits adventitious and axillary shoot formation.
Autopolyploid: Polyploids which originate by multiplication of the chromosome of a single speices.

B

Backcross method: In backcross method of breeding, the hybrid and the progenies in subsequent generations are repeatedly backcrossed to one of the parents.
Balanced population: Individuals in a balanced population have equal genetic contributions from each parent.
Base analogue: A chemical whose molecular structure mimics that of a DNA base. Owing to mimicry, the analogue may act as a mutagen through mispairing.

Beta particles: Beta particles consist of light particles, each carrying a single negative electrical charge.

Biotic: Adverse effects due to pests and diseases are called abiotic stresses.

Biotype: A group of organisms within a species that resemble each other but differ physiolgically from the rest of the species.

Bisexual flower: Flower that contains all the four whorls, such as sepals petals, androecium and gynoecium.

B-Line: The fertile counterpart of A-line. It is also called maintainer of A-line.

Breeding: Development of new forms or varieties of plants by crossing and selection of offspring for desirable characteristics.

Breeder seed: It is produced by the concerned breeder or sponsoring institute or and which is used for producing foundation seed. It is of 100% genetic purity.

Bulk method : In this method no selection is practiced F_5 to F_6 generation. By the end of F_6 generation most of the plants reached to the homozygosity. Individual selection is practiced from F_6 to later generations.

C

cDNA Library: A library composed of cDNAs (complementary DNAs), not necessarily representing all mRNAs.

Certified seed: Seed produced from the foundation or certified seed under the regulation of a legally constituted agency. This class of seed is used for commercial crop production.

Chromosome: The structure in the eukaryotic nucleus and in the prokaryotic cell that carries most of the DNA.

Cleistogamy: Fertilization in an unopened flower. It is a mechanism in which flower buds do not open at all.

Clone: Organism that is genetically identifical to their parents.

Codominant: A form of inheritance where the heterozygote can be distinguished from homozygotes.

Colchicine: It is an alkaloid which is obtained from the seed of a plant called Colchicum autumnale. Colchicine induces polyploidy by inhibiting the formation of spindle fibers. Thus chromosomes divide without moving to oppsite poles resulting in doubling of chromosomes.

Combining ability: The ability to produce superior hybrids when crossed with other appropriate inbreds.

Convergent improvement: A method of improving two inbred lines simultaneously through backcrossing the single cross independently to both of its inbred parents, that is, (AxB) xA and (AxB) xB.

Copyright: Copyright is a legal term describing rights given to creators for their literary and artistic works.

Cytoplasmic genetic male sterility: Male sterility which is controlled by both cytoplasm and nuclear genes.

Composite variety: Developed by inter mating of selected open pollinated varieties.

Combining ability: Ability of a strain to produce superior progeny upon hybridization with other strain /strains.

Cross-pollination: Pollen grains from flowers of one plant pollinate the flowers of other plant.

Cytoplasmic male sterility : A kind of male sterility caused by cytoplasmic factors carried on mitochondria (mt DNA).

D

Dee-geo-woo-gen: A semi-dwarf variety of rice discovered in Taiwan. It was semi-dwarf, stiff strawed, erect leafed and high yielding.

Detasseling: The removal of immature tassels to facilitate controlled crossing for producing hybrid maize.

Dihybrid cross: Inheritance of two characters are studied.

Dioecious: Staminate and pistillate flowers are borne on different individuals of the same species.

Disease: An abnormal condition in the plant caused by an organism (pathogen).

Disease resistance: The ability of plants to withstand, oppose or overcome the attack of pathogens.

DNA: "Deoxyribose nucleic acid", the carrier molecule of inherited genetic information.

Domestication: Bringing of wild species under human management.

Donor parent: The parent from which one or a few genes are transferred to the recurrent parent in backcross breeding.

Double cross: When two single crosses (AXB)X(CXD) are crossed the resulting hybrid population is known as double cross.

Drought: Condition of soil moisture deficiency.

Drought avoidance: It is due to the ability of plants to maintain favorable water balance even under stress.

Drought escapes: It is due to ability of a genotype to mature early, before occurrence of drought.

Drought resistance: It is the sum total of avoidance and tolerance. It refers to the genetic ability of plants to give good yield under moisture stress conditions.

Dwarfness: The genetically controlled reduction in plant height. The Green Revolution wheat and rice varieties were based on dwarfing genes.

E

Electrophoresis: A laboratory method that uses electric current to separate DNA, RNA, or protein fragments by size.

Emasculation: The removal of anthers from a flower without damaging the stigma.

EMS: Ethyl Methane Sulphonate. It is a chemical mutagen that causes GC to AT transition.

Embryo culture: The method of culturing mature and immature embryos in media.

Euploidy: Refers to numerical change in the entire genome.

Explant : An excised fragment of a tissue or organ used to initiate an in vitro culture.

F

F_1: The first generation of a cross between genetically unrelated parents. It is also called first filial generation.

F_2: The second filial generation obtained by self-fertilization.

Frame-shift mutation: The insertion or deletion of a nucleotide pair/pairs, causing a disruption of the translational reading frame.

Foundation seed: It is produced from breeder seed and maintained with specific genetic identity and purity.

Full-sib selection: Individuals having both the parents common are known as full sibs.

G

Gamete: The haploid cell produced by meiosis. The male gamete is the pollen grain, while the female gamete is the egg cell.

Gamma radiation: The electromagnetic radiation has highest frequency and energy. Electromagnetic radiation emitted by an unstable nucleus when changing to a stable nucleus is called gamma radiation.

GCA: It is the average performance of a strain or genotype in a series of crosses or hybrid combinations.

Glume: Bract of the spikelet occurring in pairs.

Gene: The unit of heredity, transmitted from generation to generation during reproduction.

Genetic diversity: The variety of different types of genes in a species or population.

Genetic male sterility: Pollen sterility which is caused by nuclear gene. The male sterile condition is ordinarily monogenic recessive.

Genetic resources: Sum total of all the genes in a crop species.

Genetically modified organisms: The organisms whose genetic make-up has been altered by the insertion or deletion of small fragments of DNA from the same or another species in order to create or enhance desirable characteristics.

Genome: All of the genetic material in the full set of haploid chromosomes in an individual.

Genotype x Environment (GxE): The interaction of a plant's genotype with the environment in which it is grown that contributes to its performance.

Genotype: The genetic makeup of a cell, organism, or individual, often with regard to a particular gene of interest.

Germinal mutations: Mutations occur in the reproductive cells or germ cell that produce gametes.

Germplasm: A collection of individual plants that contains a species' genetic variation.

H

Half-sib family selection: Individuals having one parent in common are called half-sibs.

Haploid : Having a single set of chromosomes, for example as in a gamete.

Heterogeneous population : Genetically dissimilar plants constitute heterogeneous population.

Heteroploidy: Any change in chromosome number from the diploid state.

Heterosis: Superiority of F_1 over their parents in terms of yield and related traits.

Heterozygous: Having different alleles at a particular locus.

Homogenous population: Genetically similar plants constitute homogeneous population.

Homozygous: An individual having two copies of the same allele at a given gene.

Hybrid varieties: Hybrid varieties are the first generation (F1) from crosses between two pure lines, inbred, open pollinated varieties, clones or other populations that are genetically dissimilar.

Hypersensitivity: Immediately after the infection several host cells surrounding the point of infection are so sensitive that they will die.

I

Ideotype breeding: Ideotype breeding can be defined as a method of crop improvement which is use to enhance genetic yield potential through genetic manipulation of individual plant character.

Immunity: When the host does not show the symptoms of disease.

***In Vitro*:** When something is performed in vitro, outside of a living organism.
***In Vivo*:** In a living cell or organism.
Inbred line: It is a nearly homozygous line obtained by continuous selfing in a cross-pollinated crop followed by selection.
Inbreeding depression: The loss or decrease in vigor due to continuous inbreeding.
Inflorescence: A cluster of flowers; the arrangement and mode of development of the flowers on a floral axis.
In-situ conservation: Conservation of germplasm under natural conditions.
Industrial property: which includes inventions (patents), trademarks, industrial designs, and geographic indications of source.
Intellectual property rights: Intellectual property rights (IPR) are the rights given to people over the creation of their minds. It includes, creations of the mind: inventions, literary and artistic works, and symbols, names, images, and designs used in commerce.
Introgression: Movement of a gene or locus from one species into another by hybridization.
Irradiation: Exposure of plants, plant parts, seeds, etc. to any kind of radiation to increase mutation rates.
Isogenic lines: Lines which are identical in all respect except for one gene. Such lines are produced usually through back cross method.

L

Landraces: Primitive cultivars that has been developed by adaptation to its local environment; usually heterogeneous varieties.
Linkage drag: The negative association of non-targeted genes that are inherited with a gene of interest due to linkage.
Lethal mutations: The mutations which result in the death of the individuals in which they occur.

M

Macro-mutation: Mutations with distinct morphological changes in phenotype.
Male sterility: Male sterility is characterized by non-functional pollen grains, while female gametes function normally.
Mapping: The process of identifying the location of a gene or DNA segment along a chromosome.
Mass selection: In mass selection, individual plants are selected on the basis of their phenotype, and their open-pollinated seed from them is bulked together to raise the next generation.
Meristem culture: Cultivation of axiliary or apical shoot meristems, particularly (shoot apical meristem).
Minikit trials: The minikit trials are conducted in the farmer's field along with adoptive trials.
Modern cultivars: The currently cultivated high yielding varieties are referred to as modern cultivars.
Molecular marker: An identifiable DNA sequence on a chromosome. A marker can be a gene, part of a gene, or a sequence in a non-gene region.
Monoadelphous: When filaments are united in a one bundle but anthers remain free.
Monoecious: Plant, which have separate male and female reproductive organs on the same individual.
mRNA: .mRNA is a single-stranded nucleic acid similar to DNA, but having a ribose sugar rather than deoxyribose sugar and a uracil rather than thymine as one of the bases. mRNA is usually transcribed into protein.

Multiline varieties: They are mixtures of several purelines of similar height, flowering and maturity dates, seed colour and agronomic characteristics, but having different genes for disease resistance.

Mutagen efficiency: A ratio of specific desirable mutagenic change(s) to the undesired effects such as plant damage, sterility, lethality, and the like.

Mutagenesis: Treatment of plants or plant parts with a mutagen to increase mutation rates.

Mutant: The product of mutations.

Mutation breeding: A system of breeding in which plants/plant parts/seeds or any kinds of propagules are treated with any kind of mutagen accompanied by selection for desired types in succeeding generations.

Mutation: Sudden heritable change in the genotype of an organism that is not the result of recombination.

N

Nullisomic: An individual lacking one pair of chromosome from a diploid cell (2n-2).

Nucleus seed: It is handful of seed maintained by concerned breeder for further multiplication.

O

Obsolete cultivars: These are the varieties developed by systematic breeding effort which were popular earlier and now have been replaced by new varieties.

Oligogenic inheritance: The disease resistance is governed by one or few major genes and resistance is generally dominant to the susceptible reaction.

Orthodox seeds: Seeds of this type can be dried to low moisture content of 5% and stored at a low temperature without losing their viability are known as orthodox seeds.

Ovary culture: Culture of unfertilized ovaries to obtain haploid plants from egg cell or other haploid cells of embryo sac.

P

Patent: A patent is a protection given to a patentee for an invention for a fixed period of time by the Government.

Pathogen: An organism that produces the disease.

Pathogenicity: The ability of a pathogen to infect a host strain.

Pedigree method: Superior single plants are selected from segregating generations and a record is maintained of the parent-progeny relationships

Phenotype: The visible appearance of an organism.

Physiological race: Strains of a single pathogen species with identical or similar morphology but differ in pathogenic capabilities.

Plant breeding: Plant breeding is an art and science" and technology of improving the genetic makeup of plants in relation to their economic use for the mankind.

Poly cross: It is the progeny of a line produced through random pollination by a number of selected lines.

Polygenic inheritance: In this type the disease resistance is governed by many genes with small effects and a continuous variation for disease reaction is produced.

Polyploidy: A cell containing more than two sets of chromosomes.

Panicle: An inflorescence in which the lateral branches arising from the peduncle produce flower-bearing branches instead of single flowers.

Parthenogenesis: Development of embryo from egg cell without fertilization known as parthenogenesis.

Pure line Progeny of single homozygous self fertilized plant.

Q

QTL: Quantitative Trait Loci. This is a location of a chromosome region which contributes significantly to the overall phenotypes of a quantitative trait.

Qualitative character: A trait in which variation is discontinuous, such that phenotypes can be classified into discrete categories.

Quantitative trait loci (QTL): Regions of the chromosome associated with the inheritance of polygenic traits.

Quantitative character: A trait in which variation is continuous, such that phenotypes cannot be classified into discrete categories.

R

Race specific resistance: Resistance to one, or a few, races of a disease. This type of resistance is often conferred by a single gene and is typically ineffective when new races of a disease-causing pathogen emerge. It is also known as qualitative resistance, single gene resistance, or vertical resistance.

Reciprocal recurrent selection: A form of recurrent selection that is used to improve both GCA and SCA of a population for a character using two heterozygous testers is called reciprocal recurrent selection.

Recombinant inbred line (RIL): A recombinant inbred line (RIL) is developed by continuous self pollination of F_2 generation. Generally SSD method is used for developing RILS.

Racemose: Main axis does not terminate in to a flower but it grows indefinitely and gives off flowers laterally.

Recurrent parent: The parent which is repeatedly used in backcross breeding.

Recurrent selection for GCA: A form of recurrent selection that is used to improve the gca of a population for a character and includes heterozygous tester.

Recurrent selection for SCA: A form of recurrent selection that is used to improve specific combining ability of a population for a character by using homozygous tester is known as recurrent selection for SCA.

Recurrent selection: Reselection of plants generation after generation with inter breeding of selected plants to provide better genetic recombination.

Restriction endonuclease: An enzyme that cuts single- or double-stranded DNA at a specific base sequence called a recognition site, with each restriction enzyme detecting a different sequence.

S

Segregation: Segregation can refer to the separation of genes and their respective alleles during meiosis into new daughter cells.

Self-Incompatibility: Failure of pollen to fertilize the same flower or other flower of the same plant.

Single cross: When two inbred lines or pure lines (AXB) are crossed to produce the F1 hybrid, it is known as single cross.

Single-seed descent (SSD): Single seed is randomly selected from F2 and later generations and composited to raise the next generation After 5–6 generations each line is considered inbred.

SNP: A SNP which distinguishes two sequences based on presence/absence of a single nucleotide can be used as a genetic marker.

Somatic embryogenesis: The genesis of embryos from somatic cells through tissue culture technology. Such embryos are desiccated and encapsulated to form artificial or somatic seeds.

Somatic hybridization: The production of hybrid plants through the fusion of somatic protoplasts (wall less naked cells) of two different plant species/varieties.

Somatic mutation: A mutation in a somatic or *body* cell.

Species : A group of organisms capable of interbreeding freely with each other but not with members of other species.

Spike: An inflorescence with a more or less elongated axis, along which flowers are sessile or nearly so.

Spikelet: A unit of inflorescence in the grasses, composed of the glumes, the rachilla, and the florets.

SSR: Simple Sequence Repeat and sometimes referred to as microsatellites. The "repeat" in Simple Sequence Repeat refers to a short nucleotide repeat, typically two to three nucleotides. Variations in the number of repeats forms the basis for SSR markers.

Stigma: The portion of the pistil that receives the pollen.

Strain: A group of similar individuals from a common origin.

Syngamy: Sexual reproduction; the union of male and female gametes leading eventually to zygote formation.

Sublethal mutations: They do not lead to death of all the individuals that carry them.

Synthetic varieties: Synthetic varieties are produced by crossing in all possible combinations of selected inbred lines with good general combining ability and mixing the seeds of F_1 crosses in equal quantity.

T

Trade Secrets: A trade secret consists of any valuable business information. The business secrets are not to be known by the competitor.

T-DNA: A part of the Ti plasmid that is inserted into the genome of the host plant cell.

Tester: An inbred line, single cross, three-way cross, double cross or an open-pollinated variety used to test combining ability.

Three-way cross: It is a cross between a single cross and an inbred (AXB)XC to give hybrid population.

Tissue culture: *In vitro* culture of cells, tissues, organs, etc on an artificial medium under aseptic condition.

Top cross: When an inbred is crossed with an open pollinated variety.

Totipotency: The ability of a cell or a tissue to develop into a complete adult individual after proceeding through all the stages of development.

Trademark: It is a mark used by individual, business man, company or organization to identify the products from others.

Trait: A recognizable, measurable character in a plant, which is under some genetic control.

Transgenic: An organism containing genetic material from another organism transferred by genetic engineering.

Transition: A type of nucleotide-pair substitution involving the replacement of a purine with another purine or of a pyrimidine with another pyrimidine.

Triploid: A cell having three chromosome sets (3x) or an organism composed of such cells.

Two-line system: A system for producing hybrid rice involving only two lines (A and R lines).

V

Varietal cross: When two open pollinated varieties are mated it is known as varietal cross or population cross.

Variety: A group of individuals within a species, which are distinct in form or function from other similar arrays of individuals.

Virulence: Capacity of a pathogen to incite a disease.

Vital mutations: They do not reduce the viability of the organism that carries them. These mutations are used in crop improvement programme.

W

Wide Hybridization: Hybridization between cultivated types and their wild relatives.

X

X-rays: They are electromagnetic non–particulate radiations with a wave-length of 10 A°.

Z

Zygote: The product of the fusion between one of the pollen sperm nuclei and the egg cell of the female gamete during fertilization.

1

Plant Breeding and Its Objectives

Plant breeding is the art and science of changing and improving the heredity of plants. The art of plant breeding lies in the breeder's skill in observing and selection of superior plants. The value of selection largely depends upon the skill of the breeder. The skill that a breeder develops though a close and deep study of the concerned crop species. Before the rediscovery of the Mendelian Laws in 1900 AD, the plant breeding is as an art. The modern age of plant breeding began in the early part of the twentieth century, after Mendel's work was rediscovered. Today plant breeding is a specialized technology based on genetics. It is now clearly understood that within a given environment, crop improvement has to be achieved through superior heredity. Plant breeding developed into a science as knowledge progressed towards the genetics, cytogenetic, physiology, pathology, biochemistry, biotechnology and biometry. The fundamental of plant breeding was based on recognition of the genes. The genes were identified by their effects on the visible expression of plant traits such as whether a plant was a dwarf or tall, or the flower color was white or pink. Through systematic hybridization, a desirable gene could be transferred from one to another cultivar. More recently the science of molecular genetics proposes to advance plant breeding. Molecular genetics was ushered in with the description of chemical structure of deoxy-ribonucleic acid (DNA), the material that constitutes the genes. DNA carries the instructions for synthesis of specific proteins that determine the visible expression of particular characters. The new technology does not supplant nor diminish the importance or need for traditional selection and crossing procedures in the breeding of improved cultivars. The performance of today's cultivars reached at high levels through refined selection and hybridization breeding procedure.

The early developments in plant breeding that took place are based on archaeological evidence. The first efforts to grow wild plants probably occurred in Southeast Asia culminated in the domestication of some of the many plants with which humans experimented in 13000 BC. By 10000 BC, advanced knowledge existed of such plants as rice (*Oryza sativa*), several legumes such as pea (*Pisum sativum*), runner bean or broad bean (*Phaseolus or Vicia* spp.) and possibly soybean (*Glycine max*). At this point crops became domesticated.

Although the beginning of civilization in China and Southeast Asia are obscure and fragmented. The number of successful species domesticated indicates considerable success in plant breeding. The third site of independent agricultural development and plant breeding was in south-central Mexico between 6700 and 5000 BC. Squash and avocados were domesticated first and a number of species of maize. Domestication in the Middle East and China included the self-fertilized crop like cereals and soybean.

Hybridization is an important tool in plant breeding to increase the genetic variability. The hybridization for crop improvement was scarcely used until after 1900. In cereals and pseudo-cereals with small inconspicuous flowers, such as amaranthus, a little hybridization occurred. On the other hand, early agriculturists recognized human and animal sexuality and that mating superior animals was the basis for improvement. Camerarius was the first botanist to discover that pollen produced on male flowers is indispensable for fertilization and seed development on female plants. Camerarius conducted his experiment on the dioecious plants of hemp, spinach and maize.

History of Plant Breeding

As early as 700BC Babylonians and Assyrians artificially pollinated the date palm. During 17th Century several varieties of the Heading Lettuce developed in France. Thomas Fairchild produced first artificial hybrid in 1717. Joseph Kolreuter a German Scientist, made crosses in Solanum and studied the progeny in detail. Thomas Andrew Knight (1759-1835) produced several new fruit varieties by using artificial hybridization. Le Coutier a farmer published his results of selection of wheat in 1843 and concluded that progenies from single plants are more uniform. Patrick Shireff a Scotsman practiced individual plant selection in oat and Wheat and produced desirable new varieties. Vilmorin (1857) proposed individual plant selection based on progeny testing known as "Vilmorin Principles of Progeny Testing". He demonstrate difference of selection in self and cross pollinated crops specially *Beta vulgaris* and *Triticum aestivum.* In 1903 Johannsen based on his studies in *Phaseolus vulgaris* proposed pure line theory which states that pure line is progeny of single self fertilized homozygous plant. G.H. Shull work on Maize and studied the effect of the inbreeding and cross breeding gave the concept of heterosis. Alphanso de condolle in1882 give history and origin of cultivated plants. N.I. Vivilov in 1925 proposed eight centers of origin of crop plants. During 1960 Norman Borlaug the Noble Laureate developed Mexican semi dwarf wheat variety. The dwarfing gene was isolated from wheat variety Norin 10. Semi Dwarf Mexican wheat variety introduced in India by Dr. M.S. Swaminathan and number of high yielding wheat varieties like Kalyan Sona , Sharbati Sonora and others were developed. Development of the semi-dwarf varieties of rice revolutionized the rice cultivation. In India introduction of the TN.1 variety developed by Taiwan and IR-8 variety developed by IRRI, Philippines were introduced in 1966. C.A. Barber and T.S.Venkatraman at Sugarcane Breeding Institute, Coimbatore developed sugarcane varieties, higher in yield and sugar content. Significant contribution made in the history of plant breeding is presented in Table 1.1.

Table 1.1: Land Marks in Plant Breeding

Year	Scientist	Contribution
700 BC	Babylonians and Assyrians	Pollinated date palm artificially.
1474AD	–	The first law on patent was passed in Venice, which gave monopoly rights to artisans for their inventions.
1623	–	The House of Commons of U.K. passed the Act of proprietorship.
1682	Grew	Described stamens as male reproductive part.
1694	Camerarious	First experimental report on sexual reproduction in plants(maize)and the presence of male reproductive parts.
1717	Thomas Fairchild	Produced first interspecific hybrid i.e., Fairchild's mule, crossing between Sweet William and Carnation.
1727	Vilmorins	Established first plant breeding company in France.
1744-1829	J.B. Lamarck	Proposed the theory of inheritance of acquired characters called as Lamarckism.
1753	Linnaeus	Published species *Plantarium.*
1763	Joseph Koelreuter	First reported male sterility in flowering plants.
1778	William Herbert	Reported the role of hybridization in producing new varieties.
1761-1766	Joseph Koelreuter	Made extensive crosses in tobacco and reported hybrid vigour in the hybrids of *Nicotiana, Dianthus, Datura and Mirablis.*
1800	Andrew Knight	Attempted artificial hybridization to develop several new fruit varieties.
1823	Andrew Knight	Reported equal contribution of male and female plants to the offspring; greater variation in F_2 generation.
1823	Amici	Described entry of pollen tubes into the ovary.
1809-1882	Charles Darwin (British botanist)	Proposed Modern theory of evolution by natural selection in book "Origin of species" (1859); "Variation in Animals and Plants under Domestication" (1868), "Cross and Self-Fertilization in Vegetable kingdom" (1867).

1843	John Le Couteur and Patric Shirreff	Gave the concept of progeny test and individual plant selection in wheat (John Le Couteur) and oats (Patric Shirreff).
1849	Von Gartner	Attempted 10,000 crosses in 700 species and 80 genera from which he got 250 hybrids.
1850	Hughes and Babcock	First reported sporophytic self-incompatibility in *crepis foetida*
1852	_	British Patent Law.
1856	_	Government of India introduced the Act of Protection of inventions; this act was based on the British Patent Law of 1852.
1856	Lois de Vilmorin	Gave the concept of progeny test and used in sugarbeet.
1857	Hallet (British)	Practiced single plant selection in wheat, oats and barley.
1857	Gregor J. Mendel	Initiated experimental work on hybridization in pea.
1865	G. J. Mendel	Presented his experimental finding entitled" Versuch uber PflanzenHybriden"(Experiments in plant hybridization) before the Natural History Society of Brunn.
1866	G. J Mendel	The research paper presented in 1865 was published in the annual proceeding of the society. IV, 1865.
1872		Patents and Designs Protection Act was passed by Government of India.
1876	Charles Darwin	Published the book" The Effect of Cross and Self Fertilization in Vegetable Kingdom"
1883	_	Protection of Inventions Act was introduced in India; it was consolidated as inventions and Design Act in 1888.
1884	Strausberger	Fertilization in angiosperm and noted the fusion of sperms and egg nuclei.
1890	Rimpau	Produced triticale from a cross between wheat and rye.
1890-1962	Sir R. A. Fisher	Developed Null Hypothesis for test of significance.
1896	Hopkins	Proposed ear-to-row method of breeding.
1898	S. G Nawaschin	Developed the double fertilization.
1900	Hugo DeVries, Carl Correns and Enrich Von Tschermak	Rediscovered the Mendel's law of Inheritance.
1901	Hugo de Vries	Developed mutation theory.
1902	Haberlandt	Postulated the Concept of Totipotency.
1903	W. L. Johannsen	Developed the concept of pure line theory in French bean.
1904	Hanning	First culture of embryo of selected crucifiers.
1905	Biffen	Demonstrated that resistance to stripe of rust of wheat was controlled by a single recessive gene.

1905-1912	E.M. East and G.H. Shull	Conducted extensive experiments on inbreeding and out breeding in maize.
1907	Lutz	Discovered heteroploidy in experimental population of gigant mutant in Oenothera; the gigas mutant was an autotetraploid.
1907	R.H.Harrison	Discovered tissue culture.
1908	C.B Davenport	Proposed dominance hypothesis of heterosis.
1908	E.M. East and G.H. Shull	Proposed over-dominance hypothesis of heterosis in maize.
1908	Gates	First report of triploidy (Oenothera)
1908	G. H. Hardy and W. Weinberg	Hardy- Weinberg law for random mating populations.
1908	Hopkins	Developed ear to row method in maize.
1908	Nilsson – Ehle	Multiple factor hypothesis; bulk method of breeding.
1909	G. H. Shull	First suggested pure line method for corn. His proposal was to use single crosses for commercial planting.
1910	A.B Bruce, F. Keeble and C. Pellew	Elaborated the dominance hypothesis of heterosis .
1910	Blakeslee	Discovered the globe mutant of *Datura stramonium*, which was subsequently demonstrated by Belling in 1920, to be trisomic.
1911	Berliner	Isolated B. thuringiensis from Larvae of Mediterranean flour moth and designed it as Bacillus Thuriengiensis
1911	Barrus	Introduced the concept of Physiological races in bean anthracnose.
1911	–	Protection of designs covered by the Indian Patent and Design Act (1911) with amendments in 1978 and amended rules in 1985.
1912	East and Hays	Advocate heterosis breeding as an alternate crop improvement strategy.
1914	G.H.Shull	First used the term heterosis as hybrid vigour.
1916	Winkler	The first autotetraploid was experimentally induced in *Solanum nigrum.*
1917	D.F Jones	Theory of dominance of linked genes.
1917	Stout	Originally coined the term self-incompatibility.
1917	MeFadden	Described wheat-rye hybrids.
1918	D.F. Jones	Idea of double cross hybrid in maize.
1918	Sir Ronald A. Fisher	Partitioned genotypic variance into additive, dominance and epistatic components.

1919	H.K. Hayes and R. J. Garber	Gave the initial idea about recurrent selection and suggested commercial utilization of synthetic varieties.
1920	E.M. East and D.F. Jones	Independently gave the idea about recurrent selection.
1920	N. I. Vavilov	Law of homologous series in inherited variation(or the law of parallelism of inherited variation),and the concept of centers of origin and center of diversity of plants.
1920	H. Winkler	Coined the term genome.
1921	Sewall Wright	Developed theory of path analysis, and system of mating.
1922	–	The first double cross hybrid maize, Burr Leaming Dent released for commercial cultivation in U.S.A.
1922	Blakeslee and coworkers	First report of a monoploid (*Datura stramonium*).
1922	Sewall Wright	Gave the formula for estimating the performance of synthetics varieties (syn2 generation).
1922	Harlan and Pope	Described back cross breeding method .
1923	Sax	First described QTL mapping.
1925	Clausen and Goodspeed	Synthesised *N. digluta.*
1925	E. M. East and A,J. Mangelsdorf	Discovered gametophytic self-incompatibility in *N. sanderae.*
1925	Liabach	Use of embryo culture to recover an interspecific hybrid (*Linumperenne* x *L. austriacum*).
1926	N. I. Vavilov	Proposed the concept of crop diversity.
1927	Davis	Developed top cross method to assess combining ability and performance of the progeny of selected lines.
1927	Richey	Proposed convergent improvement method.
1927	Jones	Proposed the double cross pattern for commercial hybrids.
1927	Muller	Discovered mutagenic action of X-rays in Drosophilla.
1928	L. J. Stadler	Described the mutation effects of X-rays in maize and barley.
1928	Karpe nchenko	Developed first intergeneric hybrid *Raphano brassica* from the cross between radish (*Raphanus sativus*) and cabbage (*Brassica oleracea*) .
1930	McFadden	Successfully transferred stem rust resistance from tetra ploid to hexaplod wheat.
1930	Briggs	Started an extensive backcrossing programme to develop bunt resistant varieties of wheat.
1932	Jenkins and Brunson	Used the top cross method to test the combining ability.

1933	Rhoades	Discovered CMS in maize.
1934	Dustin	Discovered colchicines.
1934	M. T. Jenkins	Suggested formula for prediction of double cross hybrid performance from that of the four non parental single cross data.
1935	M. T Jenkins	Early testing for GCA during inbred development.
1935	Nagaharu U	Proposed the origin of tetraploid species of Brassica using a triangle, which is popularly known as U's triangle.
1936	E. M. East	Supported over-dominance hypothesis.
1936	Mahalanobis	Suggested D^2 statistics for studying genetic diversity.
1936	Smith	Developed descriminate function analysis.
1937	J.B. Harrington	Proposed mass – pedigree method of breeding.
1937	Blackslee and Nebel (independently)	First described the chromosome doubling action of colchicines.
1938	–	Established International Organization for Plant Variety Protection (ASSINSEL).
1939	Goulden	Proposed single seed descent method for advancing segregating generations of self-pollinated crops.
1939	R.J. Gauthert, P.R. White and Nobecourt (independently)	The first continuous growing callus cultures were established from cambium tissue.
1940	Harlan and coworkers	Proposed a mating system.
1940	Harlan, Martini and Stevans	Proposed a scheme for multiple convergent crosses.
1942	Spargue and Tatum	Gave the concept of combining ability.
1942	Tysdal, Kisselbach and Westover	Technique of polycross for improvement of cross-pollinated forage crops.
1943	Mather	Idea of homozygous and heterozygous balance.
1943	Jones and Clark	Commercial exploitation of male sterility to produce hybrid seed production of onion.
1944	Jones and Davis	Use of cytoplasmic male sterility to produce hybrid seed of onion; discovered 'C' male sterility cytoplasm of onion.
1944	L. J. Stadler	Gamete selection for maize inbred improvement.
1945	F. H. Hull	Proposed recurrent selection for specific combining ability; coined the term recurrent selection and over-dominance.
1945	Richey	The concept of cumulative selection.
1946	Hayes and Coworkers	Convergent improvement of double cross maize hybrid.
1946	–	A centralized plant introduction agency was initiated in the division of Botany, IARI, New Delhi.

1948	Comstock and Robinson	Originally developed the concept of biparental mating.
1948	–	The world's first mutant variety, MA-9 of cotton was developed and released in India.
1949	Chase	Suggested chromosome doubling of maize monoploids for developing doubled homozygous lines or homozygous diploid lines .
1949	Comstock, Robinson and Harvey	Reciprocal recurrent selection.
1949	Mather	Scaling tests for metric traits.
1949	J. B. S. Haldane	Proposed the maximum likelihood method.
1950	Hughes and Babcock	Reported sporophytic system of self-incompatibility in *Crepis foetida*
1950	Moore	First time reported chemically induced male sterility in maize, used maleic hydrazide.
1950	–	First hybrid variety was produced in Japan using self-incompatibility.
1951	Kihara(Japanese)	Cytoplasmic male sterility in wheat.
1951	Gerstel	Described Sporophytic self-incompatibility in Parthenium
1952	Comstock and Robinson	Concept of genetic advance under selection
1952	Jensen	Developed the concept of multiline varieties in oats.
1953	Mather	Proposed the concept of disruptive selection.
1953	E. R. Sears	Suggested the use of nullisomics for location of genes in Chinese spring wheat.
1954	N. E. Borlaug(American)	Method for developing multiline varieties of wheat.
1954	Mac'key	Suggested modification of backcross method.
1954	Lerner	Concept of genetic homeostasis.
1955	Edwardson	In Texas CMS lines, fertility restored by two dominant genes Rf1 and Rf2 (later confirmed by Duvick, 1965).
1955	Hayes and coworkers	Utilization of back- cross method for incorporating smut resistance in three-way cross maize hybrid, 'Minhybrid 301'.
1956	Flor	Proposed the gene for gene hypothesis .
1956	Federer	Proposed augmented field plot design.
1956	E. R. Sears	Developed the procedure for transfer of rust resistance from Aegilops to Chinese spring wheat using irradiation.
1956	–	A project for Intensification of Regional Research on Cotton, Oilseeds and Millets(PIRRCOM) was initiated in order to intensity research on these crops.
1957	Anderson	Meteroglyph analysis of genetic variability.

1957	Kempthorne	Developed line x tester analysis, Partial diallel mating design. Book: "An Introduction to Genetic Statistics"
1957	_	All India Coordinated Maize Improvement Project was started with the objective of exploiting heterosis.
1957	_	The copyright law in India (The Indian copyright Act (1957) amended in (1999) are as per International standard.
1957	F. Skoog and R. A. Miller	Auxin, cytokinin ratio for root and shoot differentiation.
1958	Hayman	Proposed the six and five parameter models for generation mean analysis.
1958	Jinks and Jones	Developed three parameter models for generation mean analysis.
1958	J. M. Thoday	Elaborated the concept of disruptive selection.
1958-59	Reinert and Steward	Reported development of somatic embryos from callus clump and callus suspension of carrot.
1959	Dewey and Lu	First used path analysis for selection.
1959	Braum	Regenerated first plant from mature plant cell.
1960-1968	Dhawan	Feasibility of using composite varieties.
1960	Lonnquist	Used convergent improvement method for the genetic improvement of a single cross hybrid.
1960	G. Morel	Developed technique of clonal propagation; used knudson C medium for culturing shoot-tips and protocorm.
1960	E.C. Cocking	Produce large amounts of protoplasts using cell wall degrading enzymes cellulase, isolated from fungus *Myrothecium verrucaria.*
1961	Gardner	Proposed stratified or grid method of mass selection.
1961	Cockerham	Demonstrated that the expected genetic gain in yield potential of single cross hybrids would be at least double the gain for double cross hybrids.
1961	_	Released four maize hybrids, viz. Ganga 1, Ganga 101, Ranjeet and Deccan in India.
1961	Downy and coworkers	Isolated the first 'zero' erucic acid selection of B. napus.
1961	_	Independent division as Division of Plant Introduction in IARI.
1961	Thoday	Concept of QTL mapping was further elaborated.
1961	_	The first International Union for Protection of New Plant Varieties (UPOV) convention was signed in Paris.
1962	Kanta	*In vitro* pollination of opium poppy.

1962	Cockerham and Rawlings	Triallel and quadriallel analyses.
1963	Hayes	Developed the procedure for using backcross method for improvement of inbred lines.
1963	Leffel	Suggested the production of double cross hybrids in red clover using four self-incompatibility alleles.
1963	Suneson	Outlined the technique for production of commercial barley hybrid seed.
1963	Van der Plank	The idea of race-specific and race nonspecific, and horizontal and vertical resistance.
1964	N.E Borlaug	Semi dwarf high yielding varieties of wheat, which resulted in quantum jump in production and green revolution.
1964	Mertz, Batesland and Nelson	Opaque-2 mutation modified the amino acid. composition (enhanced lysine and tryptophan contents) of mature endosperm of maize.
1964	Lonnquist	Modified ear-to-row selection.
1964	S. Guha and S.C. Maheshwary	First time produced haploid plant in Datura through anther culture.
1964	Thompson	Production of triple cross hybrid kale based on sporophytic self- incompatibility in Brassica.
1964	Downey	First 'zero' erucic acid selection was isolated in *B. campestris.*
1964	_	The first hybrid variety of jowar, CSH 1, was released.
1965	N.E. Borlaug	Proposed the idea of multiline hybrids.
1965	R.T. Ramage	The balanced tertiary trisomics(BTT) system for GMS-based hybrid seed production in barley.
1965	Dhawan	Step-ladder method for the improvement of inbreed lines of maize.
1965	Athwal	First commercial hybrid of bajra, HB 1
1965	J.E. Grafius	First applied single seed descent (SSD) method in oats.
1965	Zhukovsky	Recognized 12 mega-gene centers of crop plant diversity.
1966	_	Rice varieties Taichung Native-1 (TN-1) developed at Taiwan and IR-8 developed at IRRI were introduced in India.
1967	_	Developed six composite varieties of maize , viz, Vijay, Kisan, Amber etc.,
1968	C M Donald	Proposed the concept of ideotype in wheat and also coined the term "Ideotype".
1968	Kearsey and Jinks	Developed triple test-cross analysis.
1968	Zhukovsky	Set forth the concept of mega centers and endemic micro centers of cultivated plants and their relatives.
1968	Takebe and coworkers	Use both pectinase and cellulase enzyme to isolate mesophyll protoplast.

1968		First low erucic acid variety, Oro was released in B. napus.
1969	Borthakur	Outlined the steps involved in the development of multiline, synthetic variety.
1970	Jensen	Originally developed the concept of diallel selective mating design .
1970	C.T. Patel	First developed cotton hybrid H4.
1972	Carlson, Smith and Dearing	First somatic hybrid by protoplast fusion in tobacco was produced .
1972	C.J. Driscoll	Discovered the XYZ system for the use of GMS for hybrid seed production in wheat.
1972	Zodak	Proposed the monocyclic and polycyclic approach for testing of horizontal resistance.
1973	Fasoulas	Developed the Honey-comb design for selecting individual plants from a genetically variable population.
1976	L. P. Yuan and coworkers	Developed world's first rice hybrid for commercial cultivation in China.
1977	D.R. Marshal	Given clean crop and dirty crop approach of multilines.
1978	Priestley	Proposed 'boom and bust cycle' for cereal varieties; Caused by the emergence of new races of air-borne fungal diseases.
1983	Lin and Poushinky	Proposed modified augmented design.
1991	_	The world's first GMS based hybrid variety, ICPH 8of pigeon pea was developed by ICRISAT for commercial cultivation in India.
1993	_	A photoperiod –sensitive cytoplasmic male sterility was first described in alloplasmic lines of wheat variety, Norin 26 carrying cytoplasm of Aegilops crassa.
1993	_	On December 29, 1993, India and other nations signed the International Convention on Biological Diversity.
2001	_	Hybrid variety PRH 10(Pusa Rice Hybrid 10) of basmati rice was released for commercial cultivation.
2001	_	The Protection of Plant Varieties and Farmer's Rights Act (2001) was passed on August 9, 2001 by the Lok Sabha.
2002	_	India has enacted the Biological Diversity Act, 2002.
2005	_	ICRISAT, Hyderabad released CMS- based pigeon pea hybrid ICPH 2671 (Pushkal) for Commercial cultivation.

Objectives of Plant Breeding

Plant breeding aims to improve the characteristics of plants so that they become more desirable agronomically and economically. The specific objectives may vary greatly depending on the crop under consideration.

1. Higher seed yield

High yield is the prime aims of any breeding program. The ultimate aim of plant breeding is to improve the yield of economic produce. It may be grain yield, fodder yield, fiber yield, tuber yield, cane yield or oil yield depending upon the crop species. Improvement in yield can be achieved by developing more efficient genotypes than the existing varieties.

2. Improving the quality

Improving the quality parameters like, grain size, shape, colour, aroma and cooking quality in rice, milling, baking quality and gluten content in wheat, malting in barley, size, shape and flavor of fruits, keeping quality of vegetables, protein content in legumes, methionine and tryptophan content in pulses and improving sulphur containing amino acids in oilseed crops.

3. Reducing toxic substance

Decreasing the HCN content in jowar plants,lathyrogen content in Lathyrus sativus (βN oxalyamine alanine BOAA), erucic acid in brassicas, cucurbitacin in cucurbits etc.

4. Disease and pest resistance

Resistant varieties against insect-pest and diseases help to increase the productivity and stabilize the productivity. Breeding for insect- pests and diseases resistant varieties, there by reducing the cost of cultivation, environmental pollution and saving beneficial insects. Breeding for resistant varieties is the easy way to combat abiotic stress. Due to insect-pest and diseases severe losses in yield and other traits take place.

5. Change in maturity duration

Earliness is the most desirable character which has several advantages. Development of early maturing varieties are suitable for sowing at different dates of planting, crop rotation, drought conditions, so that they can complete their life cycle before water deficit develops.

6. Improvement of the agronomic characters

It includes plant height, branching, tillering capacity, growth habit, erect or trailing habit etc., is often desirable. For example, dwarf ness in cereals is generally associated with lodging resistance and better fertilizer response. Tallness, high tillering and profuse branching are desirable characters in fodder crops.

7. Photo and thermo insensitivity

Development of the photo and thermo insensitive varieties of whcat, rice, maize, pearlmillet and other crops, so that they can grown in new areas.

8. Non-shattering nature

Development of varieties of non-shattering types, which causes serious loss in yield in crops like, green gram, brassicas, black gram, horse gram, isabol etc.

9. Synchronized maturity

It refers to maturity of a crop species at one time. The character is highly desirable in crops like greengram, cowpea, and cotton where several pickings are required for crop harvest. Breed the varieties that can mature at a single time.

10. Determinate growth habit

Development of varieties with determinate growth habit is suitable in crops like mung, pigeon pea, cotton etc.

11. Elimination of dormancy

Breed the varieties, free from toxic compounds to make them safe for human consumption. For example, removal of neurotoxin in Khesari–lentil (*Lathyruys sativus*) that causes paralysis, erucic acid from Brassica which is harmful for human health, and gossypol from the seed of cotton is necessary to make them fit for human consumption.

12. Wider adaptability

Suitability of a variety for general cultivation, over a wide range of environmental conditions. Adaptability is an important objective in plant breeding because it helps in stabilizing the crop production over regions and seasons.

13. Moisture stress and salt tolerance

Development of varieties tolerance to salinity, alkalinity and drought for those areas which are affected by these stresses. This would help in increasing the crop production.

Plant Breeding its Applications and Future Prospects

1. Creation of the Variation

Genetic variation can be created by domestication, germplasm collection, plant introduction, hybridization, mutation, polyploidy, somaclonal variation and genetic engineering.

2. Selection

Identification and isolation of superior plants with desirable characters and growing their progenies.

3. Evaluation

Newly selected strains/populations tested for their performance with standard variety used as check at several locations to see their performance.

4. Multiplication

Large scale seed production of certified seeds of the released varieties for commercial cultivation.

5. Distribution

Certified seeds distributed to the farmers for the commercial production.

Institutes engaged in Plant Breeding at National and International Level

1. International Rice Research Institute (IRRI), Philippines.
2. International Maize and Wheat Improvement Centre (CIMMYT), Mexico.
3. International Crops Research Institute for the Semi-Arid Tropics (ICRISAT), Hyderabad.
4. International Potato Centre (CIP), Peru.
5. International Board of Plant Genetic Resources (now International Plant Genetic Resources Institute; IBPGR, now IPGRI).
6. Central Potato Research Institute (CPRI), Kufri (Shimla).
7. Indian Agricultural Research Institute (IARI), New Delhi.

8. Sugarcane Breeding Institute (SBI), Coimbatore.
9. Jute Agriculture Research Institute (JARI), Barrackpore, West Bengal.
10. Indian Grassland and Fodder Research Institute (IGFRI), Jhansi.
11. Forest Research Institute, Dehradun.
12. Central Arid Zone Research Institute (CAZRI), Jodhpur.

Achievements of Plant Breeding

1. The Green Revolution started in 1943 when the Mexican government and the Rockefeller Foundation co-sponsored a project, the Mexican Agricultural Program, to increase food production in Mexico.The first target crop was wheat, and the goal was to increase the wheat production by a large margin. The important genotypes used by Norman Borlaug in his breeding program were the Japanese "Norin" dwarf genotypes.The first Mexican semi dwarf cultivars, "Penjamo 62" and "Pitic 62" were developed. In 1963 ICAR introduced Mexican variety and developed Sonalika and Kalyansona by hybridization programme, which were tolerance to water lodging and responsive to fertilizer doses.
2. Semi dwarf rice varieties developed in India by introducing gene Dee-Geo-Woo-Gene in 1966 from TN.1 variety developed in Taiwan and IR-8 from Philippines.
3. Noblization of Indian Canes: C.A. Barber and T.S. Venkatraman at Sugarcane Breeding Institute, Coimbatore developed sugarcane varieties higher in yield and sugar content. They transferred thick stem higher sugar content and other desirable characters from Noble cane to Indian cane.
4. Development of hybrid and synthetic varieties: In Maize Ganga safed-2, African Tall, Manjari, Deccan etc., Sorghum: CSH 1,2,3,4,5,6,7,8, 9,10,12,14,15R, Bajra : WCC-75, PHB-10, ICTP-8203, HB1 to HB4, Shradha and Saburi, Cotton-H-4,Var. Laxmi, Savitri, NH-44, Jay-Laxmi.
5. Developed first hybrid ICPH-8 in pigeon pea by using genetic male sterlity.
6. Released first hybrid H-4 in cotton

2

Centers of Origin

Centre of origin are also known as centre of diversity. N. I. Vavilov in 1926 has proposed that crop plants evolved from wild species in the areas showing diversity and termed them as primary centers of origin. Primary centers, crops moved to other areas with the movements of man. But in some areas, certain crop species show considerable diversity of forms although they did not originate there. Such areas are known as secondary centers of origin of these species. Genetic diversity is essential for crop improvement. Vavilov has suggested eight main centers of origin of crop plants.

1. Chinese Centre

This centre is considered to be one of the earliest and largest independent center of origin of cultivated plants. This centre includes mountain regions of central and western China. The endemic species listed from this centre include, soya bean, radish, turnip, pear, peach, plum, colocasia, buckwheat, opium poppy, brinjal, apricots, oranges, china tea etc.

2. Himalayan Centre

This centre is also known as the Indian centre of origin. This centre includes regions of Assam, Burma, Indo-china and Malayan Archipelago. The crops originated from this centre are rice, red gram, chick pea, pea, mung, brinjal, cucumber, sugar cane, black pepper, moth bean, rice bean, cotton, turmeric, indigo, millets etc.

3. Mediterranean Centre

This centre includes borders of Mediterranean Sea. Most of the cultivated vegetables have their origin in this region. This centre include crops like, durum wheat, emmer wheat, oat, barley, lentil, pea, grass pea, broad bean, cabbage, asparagus, pepper mint etc.

4. Abyssinian Center

This centre includes Abyssinia, Eritrea and part of Somaliland. It is the centre of origin of barley, sorghum, pearl millet, cowpea, flax, sesame castor, coffee, okra etc.

5. Central Asian Centre

This centre includes north-west India, Afghanistan, Uzbekistan and western China. The species originated from this centre include bread wheat, club wheat, sesame, linseed, muskmelon, carrot, onion, garlic, apricot, grape, hemp, cotton etc.

6. Asia Minor Centre

This centre covers near East Asian regions like Iran and Turkmenistan. The endemic species listed from this centre include wheat, rye, pomegranate, almond, fig, cherry, walnut, alfa alfa, persian clover etc.

7. Central American Centre

This centre includes southern parts of Mexico, Costa Rica, Guatemala and Honduras region. The species originated from this centre include maize, rajma, lima bean, melon, pumpkin, sweet potato, arrow root, chilly, cotton, papaya, guava, avocado etc.

8. South American Centre

This centre includes Peruvian regions, Islands of southern Chile, Brazil and Paraguay regions. This centre include potato, sweet potato, lima bean, tomato, papaya, tobacco, quinine, cassava, rubber, ground nut, cocoa, pineapple etc.

9. Types of Centre of Origin

1. Primary centers of origin

Primary centers are original home of the crop plants. These are the regions of vast genetic diversity of crop plants. These are generally uncultivated areas like, mountains, hills, river valleys, forests, etc. They have the following main features:

i) Have more genetic diversity.

ii) Have more number of dominant genes.

iii) Generally possess wild characters.

iv) Showed less crossing over.

v) Operates natural selection.

2. Secondary centers of origin

These are the areas where certain crop species show considerable diversity although they did not originate there. These are generally the cultivated areas and have following main features.

i) Have lesser genetic diversity in comparison to primary centers.
ii) Exhibits large number of recessive genes.
iii) Exhibits more crossing over.
iv) Operates both natural and artificial selections.

3

Crop Genetic Resources and Introduction

Germplasm is the sum total of all the genes and their alleles present in a crop and its related species. Germplasm is also termed as genetic resources. It may be release as a variety, subjected to selection for developing a variety, and may be used as a parent in hybridization programme. They may be obtained from gene banks, center of diversity, gene sanctuary, farmer' s field and seed companies.

Types of Germplasm

1. Land races

These are primitive varieties/cultivars. They are storehouse of genetic variability. Land races were first collected and studies by N.I.Vavilov in rice. They are the source of development of varieties in many valuable crops.

2. Obsolete varieties

The improved varieties of recent past are called as obsolete varieties. These varieties were breed by systematic breeding methods, were once commercially cultivated, but are no more grown. For example, wheat varieties K65, Ph591, many NP series etc.

3. Varieties in cultivation

The varieties in cultivation are the easiest to use in breeding programmes. They are good sources of genes for yield and quality.

4. Wild forms and wild relatives

Wild forms are the wild species from which crop species were directly derived. They are easy to cross with the concerned crop species.

Germplasm Conservation

It refers to the protection of genetic diversity from genetic erosion. The germplasm has to be maintained in such a stat that it can be planted directly in the field.

1. *In situ* germplasm conservation

Conservation of germplasm under natural habitat or in the area where they grows naturally is known as *in situ* germplasm conservation. This is achieve by protecting this area from human interferences; such an area are called as natural park, biosphere reserve or gene sanctuary.

2. *Ex situ* germplasm conservation

Conservation of germplasm away from its natural habit is called *ex situ* germplasm conservation". This is most practical method of conservation. The germplasm are conserved in seed gene banks, plant or field gene banks, shoot-tip gene banks, cell and organ gene banks, DNA gene banks etc.

Types of Seed Collection

1. Base collections

Plant materials which are stored for long time is known as base collection. These consist of all the accessions present in the germplasm of a crop which are stored at about -20°C with 5% moisture content. They are distributed only for regeneration. High quality orthodox seeds can maintain good viability up to 100 years.

2. Active collections

The plant materials that are stored for medium term storage (10-15 years). The accessions in an active are stored at temperatures below 15°C and the seed moisture is kept at 5%.

3. Working collections

The germplasm which are stored for short period. Such accessions being actively used in crop improvement programmes. Their seeds are stored for 3-5 years at less than 15°C with 10% moisture These collections are maintained by the breeders.

Plant Introduction

In India, Portuguese introduced new-world crops, such as maize, groundnut, chilli, potato, sweet potato, guava, custard apple, pineapple, cashewnut and tobacco. British East India Company introduced - tea, litchi and loquat from China and cabbage, cauliflower and other winter vegetables. Introduction is of two types:

1. Primary introduction

When the new material is well suited to new environment then it is released as a variety for commercial cultivation without any change of genotype. For examples, - dwarf wheat varieties like 'Sonora-64' Lerma rojo and dwarf rice varieties like "Taichung Native 1, 'IR-8 and IR-36.

2. Secondary introduction

When introduced material is subjected to selection or used in hybridization programme with local varieties to get the improved varieties with some new characters. For example, 'Kalyan Sona' and 'Sonalika' varieties of wheat introduced from CIMMYT, Mexico.

Procedure of Plant Introduction

1. Procurement of plants or germplasm

Requisition for introduction of new crop plant or new varieties should be submitted to NBPGR (National Bureau of Plant Genetic Resources) within the country or to IBPGR (International Bureau of Plant Genetic Resources). The material may be obtained on an exchange basis from friendly countries or the material can also be purchased or obtained as free gift from individuals or organizations.

2. Packaging and dispatch

Parts of the plant for propagation of that species is known as propagule. Propagule may be seeds, tubers, runners, suckers, stolons, bulbs, root cuttings and buds. Propagule are packed carefully and dispatched.

3. Entry and plant quarantine

On receipt of the material, the entry inspection is done by the country for other contaminants and the presence of insect, diseases and nematodes are checked. The materials are treated with insecticides, fungicides or nematicides and then released.

4. Cataloguing

Introduced specimen is given a number regarding species, variety, place of origin and the data are recorded.

5. Evaluation

New material is tested at different locations or places to see their performance.

6. Multiplication and distribution

Promising introduced materials are propagated and then released as varieties for commercial cultivation after necessary trials.

4

Apomixis in Plant Breeding

Apomixis is an asexual method of reproduction through the seed. Apomictic reproduction is widespread and found in over 300 species of plants in 30 families. Interest in apomixis as a reproductive mechanism has increased as its potential and role are being recognized in genome preservation, polyploid build up, propagation of derivatives of wide crosses, permanent fixation of hybrid vigour, evolution, speciation and plant breeding .

Winkler (1906) defined apomixis as "the substitution for sexual reproduction of another asexual reproductive process that does not involve nuclear or cellular fusion." One of the chief barriers to an understanding of the apomictic phenomenon has been the highly complex terminology as well as the differences of opinion among specialists as to the correct usage of these terms, probably because the phenomenon by itself is complex and in many ways quite different from that of amphimixis. Convincing evidence for the presence of apomixis can be obtained by tracing the steps that are involved in embryonic development. It should be amply clear that as the prerequisite for apomixis, fertilization does not take place at all in any form. Thus there is no normal development of embryo from the egg cell. However, the embryo may develop from an unfertilized egg cell other than the egg cell within the embryo sac.

Apomixis is widely distributed among higher plants. More than 300 species belonging to 35 families are apomictic. It is most common in Gramineae, Compositae, Rosaceae and Rutaceae. Among the major cereals like maize, wheat and pearl millet have apomictic relatives.

Types of Apomixis

1. Obligate apomixis

A plant which reproduce can only have apomictic embryos.

2. Facultative apomixis

A plant which has the potential to reproduce from both apomictic and normal embryos is known as facultative apomixis.

Classification of Apomixis

1. Recurrent apomixis

An embryo sac develops from megaspore mother cell (MMC) where meiosis is disturbed (sporogenesis failed) or from some adjoining cell (MMC disintegrates). Consequently the egg cell is diploid. The embryo subsequently develops directly from the diploid egg cell without fertilization. Somatic apospory, diploid parthenogenesis and diploid apogamy are recurrent apomixis. However, diploid parthenogenesis/apogamy occur only in aposporic (Somatic) embryo sacs, Therefore, it is the somatic or diploid apospory that constitutes the recurrent apomixis, Such apomixis occurs in some species of Crepis, Taraxacium, Poa (blue grass) and Allium (onion) without the stimulus of pollination and in Parthenium, Riubus (raspberry), Malus (apple)and Rudbeckia where pollination (not fertilization) appears to be necessary either to stimulate embryo development or to produce a viable endosperm.

2. Non-recurrent apomixis

An embryo sac arises directly from normal egg cell (n) without fertilization. Since an egg cell is haploid, the resulting embryo will also be haploid. Generative apospory and related processes, namely, haploid parthenogenesis and haploid apogamy and androgamy fall in this category. Such types of apomixis are of rare occurrence. It is useful for development of inbred lines.

3. Adventive embryony

Embryos arise from a cell or a group of cells either in the nucellus or in the integuments, eg., citrus species. The term was first used by Strasburger in 1878. Since it takes place outside the embryo sac, it is not grouped with recurrent apomixis, though this is regenerated with great accuracy. In addition such embryos, regular embryo within the embryo sac may also develop simultaneously, thus giving rise to poly-embryony condition as in citrus and opuntia.

4. Vegetative apomixis

In some cases like Poa Bulbosa and some Allium, Agave and Grass species, Gegetative buds or Bulbils, instead of flowers are produced in the inflorescence. They can also be reproduced without difficulty. However, Russian workers do not group this type of vegetative reproduction with apomixis.

5. Apospory

When embryo develop directly from the diploid egg cell without fertilization is called apospory. The term apospory was coined by De Very and Bower in 1886. It is of two types:

i) **Generative or haploid apospory**: If the embryo sac develops from megeaspoes (n) the process is called generative or haploid apospory. Since it cannot regenerate as it is haploid and fertilization fails, the process gives rise to non-recurrent apomicts.

ii) **Somatic or diploid apospory**: When diploid embryo sac is formed in nucleus or other cells, the process is termed as somatic or diploid apospory. Since it regenerates without fertilization, it is recurrent apomicts.

6. Apogamy

This mainly relates to egg cell but also from any one of the synergids or antipodal cells within the embryo sac without fertilization. The term apogamy was given by Winkler in 1908. It has been reported in Allium, Iris and other plant species. It is also of two types:

i) **Haploid apogamy:** Embryo develops in a haploid embryo sac.

ii) **Diploid apogamy:** Embryo develops in a diploid embryo sac.

7. Androgamy

Development of embryo from one of the male gametes itself, inside or outside the embryo sac is known as androgamy.

8. Parthenogenesis

This refers to the development of embryo from egg cell without fertilization for example, in some cases in corn, wheat, tobacco, barley, bulbosum and cotton. The term parthenogenesis was discovered by Owen in 1849.This is also of two kinds:

i) **Haploid parthenogenesis**: The embryo develops from egg cell without fertilization in a haploid embryo sac produced by generative apospory. It is non-recurrent in nature.

ii) **Diploid parthenogenesis**: The embryo develops from egg cell without fertilization in a diploid embryo sac arising from somatic apospory. It is very useful in conserving heterosis. It is recurrent type.

9. Automixis

Fusion of two haploid nuclei in a meiotic embryo sac or restitution in the second division of meiosis also results in diploid reproductive cells. Such diploid cells will be completely homozygous and as such are useful in the practical exploitation of apomixis.

Advantages of Apomixis in Plant Breeding

The two sexual processes, self and cross fertilization, followed by segregation, tend to alter the genetic composition of plants that reproduced through amphimixis. Inbreeding and uncontrolled out breeding also tend to break superiority in such plants. On the contrary, apomicts tend to conserve the genetic structure of their carriers. They are also capable of maintaining heterozygote advantages generation after generation. Therefore, such a mechanism might offer a great advantage in plant breeding where genetic uniformity maintained over generations for both homozygosity and heterozygosity is the choicest goal. Additionally, apomixis may also affect an efficient exploitation of maternal influence, if any, reflecting in the resultant progenies, early or delayed, because it causes the perpetuation of only maternal individuals and maternal properties due o prohibition of fertilization. Maternal effects are most common in horticultural crops, particularly fruit trees and ornamental plants. Thus, in short, the benefits of apomixis so far as their utility in plant breeding is concerned are:

1. Rapid multiplication of genetically uniform individuals can be achieved without any risk of segregation.
2. Heterosis or hybrid vigour can permanently be fixed in crop plants; thus no problem for recurring seed production of F_1 hybrids.
3. Efficient exploitation of maternal effect, if present, is possible from generation to generation.
4. Homozygous inbred lines in corn can be rapidly developed and they produce sectors of diploid tissues and occasional fertile gametes and seeds.
5. Intermediate fixation of heterozygous genotypes including those made through wide crosses, allowing single plant evaluation and making possibilities for new breeding strategies and methods in sexual and vegetatively propagated crops.
6. Plant breeding could become readily responsive to micro environments cropping conditions, pathogen populations and markets stimulates diverse strategies for agro-ecosystem management and optimization.
7. Preparation of very large numbers of hybrid cultivars from crop species into which the trait is introduced.
8. Propagation of hybrid seed directly by the farmer without the need for inbreds or male steriles and without recourse to frequent seed purchases.
9. True seed propagation of traditionally vegetatively propagated crops, concomitant elimination, or reduction of disease, substantial increase in germplasm flows and potential expansion of growing regions.

10. Substantial increase in yield in some crops due to increased photosynthate availability through elimination of male flowers or flower parts.

Benefit of Apomixis

i) In vegetatively reproduced plants (e.g., potato), the main applications of apomixis are the avoidance of phytosanitary threats and the spanning of unfavorable seasons.
ii) Apomixis would greatly reduce problems due to pathogen or insect epidemics.
iii) Fixation of heterosis through apomoxis .
iv) Apomixis would lower seed production costs for industry and breeding companies and seed costs for farmers.

The Ideal Apomictic System

1. All the progeny of plants should be apomictic, so that progeny have the same genotype as the maternal parent.
2. The apomictic genotype should preferably be fully male fertile and self-incompatible and reproduce via pseudogamy.
3. In case of diplospory, chromosomes should not pair or recombine during first meiotic division. Which may give rise to variation among the progeny.
4. Apomixis should be dominant over sexual reproduction. Usually, apomixis is governed by two or more genes.
5. Expression of apomixis should be little affected by the environment.

Development of Apomictic Lines

Apomictic lines can be developed by the following different approaches:

1. **Gene transfer from wild species**: Genes can be transferred into a crop species from a related wild species, e.g., from Tripsacum dactyloides into maize and from Pennisetum orientate into pearlmillet.

2. **Induced mutations**: This approach aims developing apomictic forms in normally sexually reproducing species by utilizing induced or even spontaneous mutations.

Apomixis in Plant Breeding

1. In some cases, hybrid vigour may be conserved for many generation by using recurrent apomixis.
2. Maintenance of superior genotypes.
3. Apomixis is useful in maintaining the characteristics of mother plant from generation after generation.

4. Apomixis is an effective means for rapid production of pureline.
5. Rapid multiplication of genetically uniform individuals can be achieved without risk of segregation.

Vybrids

Heterozygous F_1 individuals with facultative apomixis produce a population containing two types of offspring, those with F_1 genotypes derived through apomixis and those with F_2 genotype originating through sexuality. The frequency of F_1 genotype will be equal to the frequency of apomixis. If next generations is derived through F_1, then the population will remain same. If the procedure is repeated, it is possible to maintain heterozygotic vigour in the subsequent generation and to keep the mean constant. Such populations are known as vybrids. Vybrids are the progeny obtained from crossing of two facultative apomicts which reproduce through faculative apomixis. They are superior varieties with yield levels and remain constant, if they are perpetuated through F_2, looking plants in each generation.

Procedure for the Production of Vybrids

The following steps are involved in the production of agronomically desirable vybrids:

1. Select desirable genotypes and cross them with one or more facultative apomicts with the latter as females.
2. Grow the F_1 generation. Build up B_2 and F_1 generations of heterotic$F_{1.}$
3. Grow F_1 ,F_2 and B_2 generations side by side. Select vigorous F_1-like plants from either the F_2 or B_2. Vigorous plants occur in greater frequency in F_2 but cross sterile occur in greater frequency in B_2. Tag the plants at the time of anthesis . Test for cross sterility by crossing a portion of the ear to other pollen parent. Self part of the ear. At maturity, harvest only the selfed seeds of the cross sterile.
4. Grow the progeny of cross sterile. Make crosses between desirable plants from different crosses
5. The seed from such crosses gives both sexual as well as facultative apomicts. Reject completely segregating sexual. Select only vybrids that have high frequency of F_1 like vigorous plants.
6. Yield test of good vybrids. Good vybrids should have a high frequency of apomixis, a high degree of yield heterosis and a minimum of negative transgressive segregation.

5

Male Sterility and It's Role in Agriculture

Male sterility refers to a condition in which pollen is either absent or non functional or unable to produce functionally viable pollen in flowering plants. Since it was first reported by Kolreuter as early in 1763 that anther abortion occurs naturally in some plant species or in their inter specific crosses, its occurrence was observed in several species. In the 19th century, Gartner (1844) and Herbert (1847) reported the occasional recurrence of anthers rudimentariness or abortion in many ornamental and vegetable crops in normal populations. During the early 20th century, a large number of species were shown to possess male sterility, for example, *Suturela hortensis*, *Lathyrus odoratus, Merculalis annua*, *Solanum tuberosum*, *Vitis vinifera*, *Linnum usitatissimum, Zea mays*, *Antirrhinum majus, Prunus persicaer* and *Rubus idaeus*. Following the rediscovery of Mendel's Laws of Inheritance, study of the genetic main basis of inheritance of male sterility attracted the attention of several workers. The main impetus for the study of inheritance of male sterility came from the elucidation of heterosis in maize by G. H. Shull, in 1908 and its exploitation through the use of male sterility by Jones and Clark (1943). The male sterility based scheme of hybrid seed production in corn opened up a new era for the hybrid corn industry as it could overcome the necessity of detasseling of female (parent) lines in the hybrid seed production process.

The male sterility may arise due to malformation of male sex organs, lack of normal anther sac, and inability of anthers to release mature pollens, lack of development of normal microspore.

Features of Male Sterility

1. Promoting allogamy.
2. Leads to heterozygosity.
3. Occurring naturally by spontaneous mutations as well as induce artificially by chemical or physical mutagens.
4. Governed by nuclear genes, cytoplasmic genes, or both.
5. Caused due to pollen or anther abortion.

Causes of Male Sterility

1. The absence, atrophy, -malformation of male sex organs in normal bisexual flowers.
2. The inability to develop either normal anther sac or anther tissue.
3. The inability of pollen to mature or be released from anther sacs or to reach or germinate on compatible stigma.
4. The failure to develop normal microspore or pollen.

Origin of Male Sterility

1. Spontaneous mutations.
2. The ultimate source of genetic variation, although mostly detrimental to the immediate survival value and evolutionary fitness of an organism, provide genetic variability that fuels evolution or offers raw material that enhances the evolutionary pace.
3. Such an example is the mutation of the self-fertility gene to self-sterility via male sterility which paves the way to encourage out breeding, accumulate heterozygosity, and utilize the advantage of hybridity for survival, perpetuation, and evolution. Although the occurrence of spontaneously arisen male sterile mutations in male fertiles is statistically a random event, spontaneous male steriles in flowering plants occur frequently .
4. Induced mutagenesis using potent and non potent mutagens has also led to their induction, but the proportion is low as compared to spontaneously arisen male steriles.

Types of Male Sterility

1. Genetic male sterility (GMS)

i) Temperature sensitive genetic male sterility (TGMS)

ii) Photoperiod sensitive genetic male sterility (PGMS)

iii) Transgenic male sterility

2. Cytoplasmic male sterility (CMS)
3. Cytoplasmic genetic male sterility (CGMS)
4. Chemically induced male sterility

1. Genetic male sterility (GMS)

Genetic male sterility, also known as nuclear male sterility, this sterility is controlled by nuclear genes whose action and expression are not influenced by the plasma or cytoplasmic genes. Therefore, inheritance pattern and expression exhibit no reciprocal differences, low environmental impact, or normal genomic influence.

Male sterility genes are usually governed by a monogenic recessive and rarely dominant genes .This type of sterility has been reported in several crops like barley, wheat, maize, cotton, sorghum ,pigeon pea, tomato, chillies etc. Male sterility alleles arise spontaneously or may be induced artificially by physical and chemical mutagens either alone or in combination. GMS has been identified in about 175 species. This system consists of two types of line, viz; A line & B line.

A line: It is male sterile line which is used as female parent in hybrid seed production. It is represented by mm (homozygous recessive).

B line: It is a male fertile line which is used to maintain male sterility in A line. It is also called maintainer line. Both A and B lines are isogenic lines with a differences only on fertility/sterility locus. When a male sterile plant (ms ms) is crossed with a male fertile (Ms Ms). The F_1 (Ms ms) is male fertile and in F_2 the ratio will be 3 male fertile and 1 male sterile i.e. 3:1 ratio (Fig.5.1). Genetic male sterile line is maintained by crossing recessive male sterile plants (mm) with heterozygous male fertile plants (Mm). Such cross gives 1:1 ratio sterile and fertile plants.

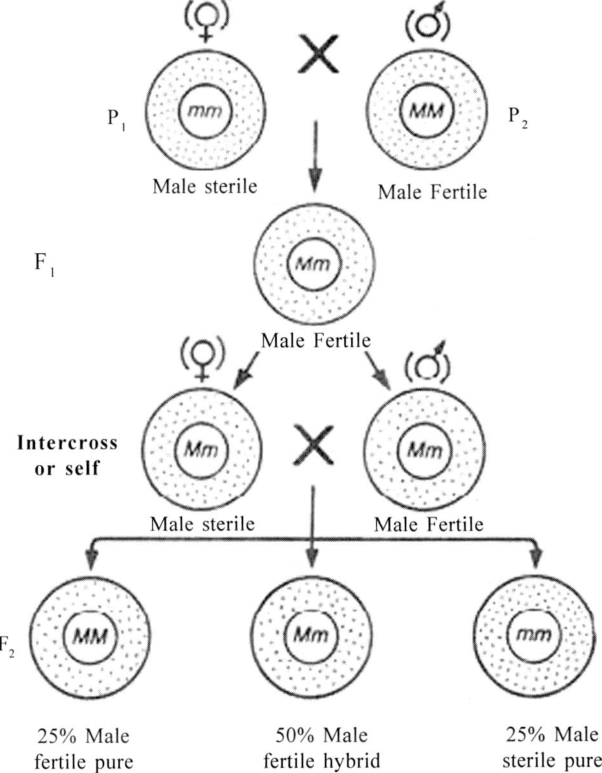

Fig. 5.1. Inheritance of male sterility

i) **Temperature sensitive genetic male sterility (TGMS)** : Thermo sensitive genetic male sterility, whose sterility/fertility alternation is conditioned by different temperature regimes, e.g., most of the TGMS lines remain male sterile at a high temperature(more than 23.3°C) and they revert back to partial fertility at a lower temperature (less than 23.3°C). TGMS sterility/fertility varies from genotypes to genotypes. The critical thermo sensitive for fertility alternation in the TGMS line varies from 15 to 25 days before heading or 5 to 15 days after panicle initiation. In India the TGMS system to be more useful than PGMS, because of non availability of longer photoperiod, which is prerequisite for PGMS system. Examples- In rice, the sterile plants produced by the male sterile gene at above 23°C temperature and become fertile below 23°C. In cotton, below 12°C sterile plants become fertile. This type of male sterility is being used in China for hybrid seed production of rice.

ii) **Photoperiod sensitive genetic male sterility (PGMS) :** Photoperiod sensitive genetic male sterility includes genetic male sterile line which respond to the photoperiod duration of day length for expression of pollen sterlity and fertility behaviors, e,g. most of the PGMS lines remain male sterile under a long day (more than 13.75h) conditions and revert back to fertility under short day (less than 13.75h) conditions. Example-in **rice**, long day condition leads to male sterility and short day condition leads to male fertility. This type male sterility is being used to developed hybrid rice in China.

iii) **Transgenic male sterility** : Transgenic is used for induction of male sterility. It's induced by technique of genetic engineering is called transgenic male sterility. Examples:-tobacco, rapeseed, maize etc. For hybrid seed production, *Barnase* and *Barstar* gene are involved. *Barnase* gene causes male sterility and *Barstar* gene restore the fertility. Both genes are produced from the same bacterium. Barnase is integrated in any plant, called A line. Gene Barstar is integrated in restorer line, called R line. Transgenic male sterility is controlled by a dominant gene called Barnase. The male sterile plants are always heterozygous. Barstar and barnase system are as follows:

- *Barnase* is extracellular RNase; barstar is inhibitor of barnase (both from bacterium *Bacillus amyloliquefaciens* which uses barnase for protection from microbial predators and *barstar* to protect itself from *barnase.*
- Fuse the *barnase* and *barstar* genes to TA29 promoter–TA29 is a plant gene that has tapetum specific expression–good for a wide range of dicot and monocot crop plants.
- Plants containing the TA29–barnase construct are male sterile; those with TA29–barstar are not affected by the transgene.

- Cross male sterile (*barnase*) with male fertile (*barstar*) to get hybrid seed, which now has both *barnase* and *barstar* expressed in tapetum and, hence, is fully fertile. *Barstar* is dominant over *barnase.*

Advantages of GMS

1. Fertility restoration in the hybrid and crossing plans are relatively easy.
2. Use in both seed propagated and vegetatively propagated crops.
3. Less area and labour are required because maintain only two lines.
4. Does not have undesirable agronomically characters.
5. Sterility is stable, reliable and replicable.
6. TGMS and PGMS mainly used in china for hybrid rice development.
7. Transgenic GMS commercially used in maize, cauliflower, tomato, wheat, chicory and closest to commercially utilization in Brassica species.

Disadvantage of GMS

1. Less stable due to GMS affected by environmental factors like temperature and day length condition.
2. Increases cost of production of hybrid seed because 50% fertile plants are removed every year.
3. In hybrid seed production programme, such males fertile plants must be eliminated before flowering and this job is very tedious, laborious and costly.

Utilization of GMS in Plant Breeding

1. GMS has been exploited commercially in few countries, like U.S.A. successfully used in castor.
2. In India pigeon pea, hybrid ICPH8, ICPH4, CoH1, CoH2, AKPH4104, AKPH2022. has been released.
3. In chillies, line Ms12 has GMS allele **ms-509** (through mutagenesis) which is used for hybrid seed production in Korea and Hungary.
4. Used limited extends to develop hybrid in tomato.

2. Cytoplasmic male sterility (CMS)

This type of sterility is controlled solely by the specific cytoplasm genes. Whose action is not influenced by nuclear or other genes. Therefore, as long as the cytoplasm of individual has these specific genes, it is male sterile cytoplasm. Pollen sterility which is controlled by cytoplasmic genes or plasma genes is called cytoplasmic male sterility. Cytoplasmic genes are most often maternally transmitted in plants. Any male sterility found in the population, if it follows maternal inheritance, is said to be cytoplasmic male sterility. The plants, which posses sterile type of cytoplasm, will not set seed unless pollinated by other fertile plants. This system consists of A line and B line. A line is male sterile line which is used as female parent in hybrid seed production. B line is a male fertile line which is used to maintain male sterility in A line called maintainer line and used as male parent. Cytoplasmic male sterile line is maintained by crossing A line with B line. CMS cannot be utilized for hybrid seed production without use of restorer line, because F_1 seed produce only male sterile plant because their cytoplasm is derived entirely from female gamete. CMS is not influenced by environmental factors like temperature and light thus it is stable. This type of sterility is used for hybrid seed production in ornamental species or in species where vegetative part is of economic value. CMS is used to develop some important varieties in sugarcane, potato, chilly etc.

Advantages

1. Highly stable because not influenced by environmental conditions such as temperature and day length.
2. Less area requires, because the breeder has to maintain only A and B line.

Disadvantages

1. Cannot be used for development of hybrids in those crops where seed is economic product.
2. It is impossible to restore fertility in hybrid as it is governed by plasma genes.
3. Sometimes, CMS line has inferior agronomic performance.

3. Cytoplasmic genetic male sterility (CGMS)

This is caused by the interaction between genetic factor (s) present in cytoplasm and the nucleus. Absence of a sterility inducing factor either in the cytoplasm or in the nucleus makes a line male fertile. Presence of certain dominant restorer genes (s) in the nucleus makes a line capable of restoring fertility in the hybrid derived from it and a CMS line (Fig. 5.2). First time reported by Jones and Davis in 1944 in onion. The cytoplasmic genetic male sterility is known in several crop plants, namely, maize, bajra, cotton, rice, sunflower, jowar, sunflower, brassica, tomato, carrot, etc. The cytoplasmic genetic male sterility system involves:

A line: It is male sterile line which is used as female parent in hybrid seed production.

B line: It is a male fertile line which is used to maintain male sterility in A line called maintainer line.

R line: Which is used to restored male sterility in male sterile plant, used as male parent in hybrid seed production.

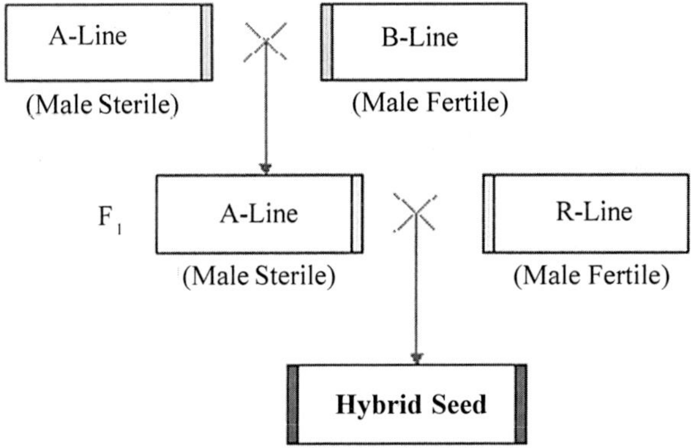

Fig. 5.2. Process of hybrid seed production

Advantages

1. Widely used for hybrid seed production in both seed propagated and vegatatively propagated species.
2. Highly stable and reliable.
3. Not affected by environmental factors.
4. In India, hybrids GTH 1, ICPH 2671, 2740 and OPH 15-3 based on CGMS in pigeonpea were developed and released.
5. CGMS is used commercially to produce hybrid seed in maize, bajra, cotton rice, sunflower, jowar etc.

Disadvantages

1. Its requires more area and labour, because the breeder has three types of materials, viz., A, B & R line.
2. Moreover, sometimes CGMS line has inferior performance.

4. Chemically induced male sterility

This non-genetic method of inducing male sterility involves the use of chemicals (gametocide) called chemically induced male sterility. Chemically induced male sterility is induced by spraying a rice variety with chemical like gametocide that can kill pollen grains of treated plants without affecting the pistil. In hybrid seed production, two parents are plated in alternate rows. One is sprayed with chemicals at appropriate growth stage and other is used as pollen source to produce the hybrid seed. The efficiency of chemical gametocides depends upon the correct doses, appropriate stage of treatment, even coverage of chemical and synchronizes flowering in the female parents. Some gametocides such as NAA, Maleic Hydrazide, Ethrel, Gibberellins, Ethidium Bromide etc. Ethidium Bromide has been used in pearlmillet and barley for induction of CMS. In china, chemically induced male sterile line has been used for hybrid seed production.

Properties of an Ideal CHA

1. Highly male or female selective.
2. Easily applicable.
3. Time of application should be flexible.
4. Should not be mutagenic.
5. Not be carried over in F_1 seeds.
6. Produce more than 95% male sterility.
7. Must cause minimum reduction in seed set.
8. Should not be hazardous to the environment.
9. Should affect the ovule fertility.
10. Application should be simple.
11. Safe for use and minimum side effect on plant growth.

6

Self-Incompatibility and Its Applications

Self incompatibility in plants is a mechanism preventing the seed set by self pollination, and is probably way in which out crossing is enforced. Other factors may cause selfed seed may not to set seed due to embryo abortion, but SI is prezygotic and prevents embryo formation. SI is, therefore prevention of fusion of fertile male and female gametes after fertilization. Koelreuter first time in 18th century reported S.I in *Verbascum phoeniceum* plants. In self-incompatible pollen grains fail to germinate on the stigma of the flower that produced them. Self incompatibility has been reported in more than 300 plant species belonging to 70 family of angiosperm. It is important out breeding device (cross pollination) for normal fruit set. It is maintained high degree of heterozygosity and can take place at any stage between pollination and fertilization.

Features of Self-Incompatibility

1. Plant does not set seed,when self-pollinated with its own functional pollen. However, set seed when it is cross pollinated,it indicated the presence of SI.
2. SI is an important out breeding device, which promote allogamy and prevent autogamy.
3. SI may be due to morphological, genetic, physiological and biochemical causes.
4. SI enhance heterozygosity and reduce homozygosity.

Types of Self-Incompatibility System

It is assumed that the two main types of SI have a common mechanism within each type.

1. Heteromorphic Self-incompatibility System
2. Homomorphic Self-incompatibility System
    i) Gametophytic self-incompatibility
    ii) Sporophytic self-incompatibility

1. Heteromorphic self-incompatibility system

When self incompatibility is associated with differences in floral morphology, it is known as heteromorphic system of self-incompatibility. This is results due to differences in the length of style and stamen. The majority of heteromophic species are distylic, such as primula have two types of flower Pin (ss) and Thrum (Ss). The flower with long styles and short stamens are called pin and flower with short styles and long stamens are called thrum. This situation is known as distyly. A few species, such as *Lythrum salicaria*; the style of a flower may be either short, long or of medium length, this condition is called Tristyly. The distyle is determined by a single gene, with two alleles. Ss produce thrum, while ss produces pin flowers. The incompatibility reaction of pollen grains is determined genotype of the plant producing them. The mating between pin and thrum plants would produce Ss and ss progeny in equal frequencies (Table 6.1). This system is of little importance in crop plants. Example., sweet potato and buckwheat.

Table 6.1. Heteromorphic self-incompatibility system

Cross	Genotypes	Results
Pin x Pin	ss x ss	Incompatible
Thrum x Thrum	Ss x Ss	Incompatible
Thrum x xPin	Ss x ss	1:1 (50% compatible and 50% incompatible)
Pin x Thrum	ss x Ss	1:1 (50% compatible and 50% incompatible)

2. Homomorphic self-incompatibility System

The majority of self incompatibility species have no differentiation in flower structure, and this is called homomorphic self incompatibility. This system of self incompatibility results due to physiological causes rather than differences in flower morphology. This system is very much important in crop plants. The incompatibility reaction of pollen is either controlled by the genotypes of the plant on which it is produced (Sporophytic control) or by its own genotypes (gametophytic control). The pollen grains do not germinate on the stigma of same flower. There are two main groups of homomorphic species, classified on the mode of action of SI mechanism and number of loci involved.

i) Gametophytic self-incompatibility

This type of self incompatibility is controlled by genetic constitution of gametes. In 1925 East and Mangelsdorf described this type of self incompatibility first time in *Nicotiana sanderae*. The incompatibility reaction of pollen is controlled the genotype of the pollen (Male gametophyte), and not by the genotype of the plant on which it is produced. Generally, incompatibility reaction is governed by a series of multiple alleles of genes designated as S_1, S_2, S_3, S_4-S_n. The gametophytic system is found in families like Solanaceae, Rosaseae, Graminae, Leguminoseae, Chenopodiaceae and Ranunculaceae. In a single gene system, this type of incompatibility gives three types of mating (Fig.6.1):

1. S_1S_2 x S_1S_2 _______ Fully incompatible
2. S_1S_2 x S_2S_3 _______ Partially (i.e., 50% of the pollen) compatible
3. S_1S_2 x S_3S_4 _______ Fully compatible

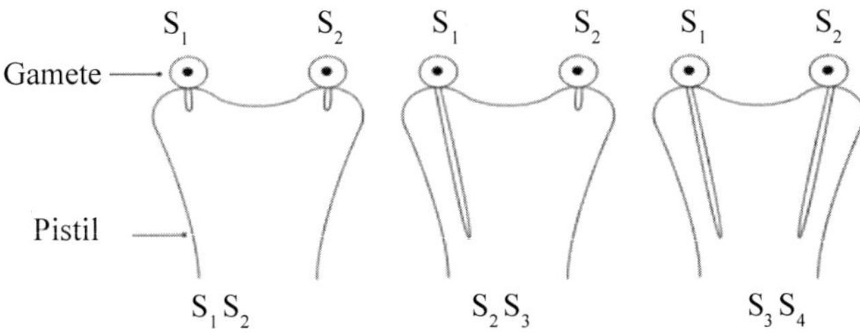

Fig. 6.1. Gametophytic self Incompatibiliti

ii) Sporophytic self-incompatibility

In this system the self-incompatibility is governed by a single gene, S, with multiple alleles. It was first reported by Hughes and Babcock in 1950 in *Crepis foetida*, and by Gerstel in *Parthenium argentatum*. Incompatibility reaction of pollen is controlled by the genotype of the plant on which the pollen is produced, and not by the genotype of the pollen. In the sporophytic system, the S alleles may exhibit dominance. For instance, the genotype of plant is S_1S_2 and S_1 is dominant to S_2 then all the pollens produced by S_1S_2 would behave as if they were S_1. Hence S_1 and S_2 pollen would be incompatible in S_1 style but compatible in S_2 style (Table 6.2 and Fig.6.2). This form of self-incompatibility has been studied intensively in members of the mustard family (Brassica), including turnips, rape, cabbage, broccoli, and cauliflower.

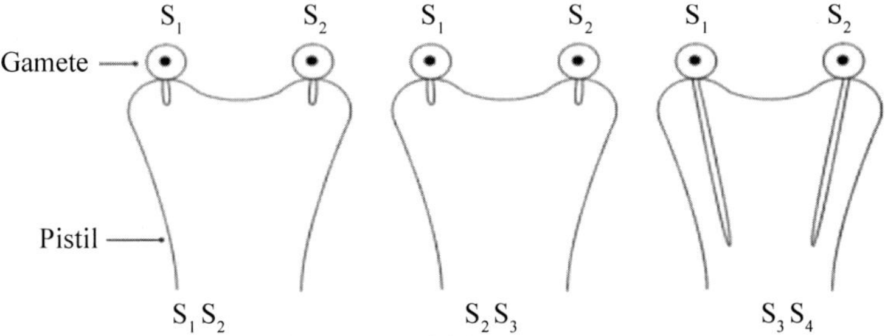

Fig. 6.2. Sporophytic self Incompatibiliti

Table 6.2. Sporophytic self-incompatibility

Cross	Compatibility
S_1S_2 X S_1S_2	Fully incompatible
S_1S_2 X S_2S_3	Fully compatible
S_1S_2X S_1S_4	Fully incompatible

Such self fertility is known as pseudo fertility and is achieved by temporarily suppressing the incompatibility reaction using one of the following techniques: Bud pollination, surgical techniques, end-of-season pollination, high temperature,. irradiation, grafting and double pollination.

Mechanism of Self-Incompatibility

The various mechanisms observed with regard to self incompatibility

1. Pollen-Stigma interaction

This condition is found just after the pollen grains reach to the stigma and generally prevents pollen from germination. Previously it was assumed that binucleate condition of pollen in gamatophytic system and trinucleate condition in sporophytic system was the reason for self incompatibility. But after that, it was found from the research that they are not the reason for S_1. Under homomorphic system of incompatibility there are differences in the stigmatic surface which prevents pollen germination. In gametophytic system the stigma surface is plumose having elongated receptive cells which is commonly known as wet stigma. The pollen grain germinates on reaching the stigma and incompatibility reaction occurs at a later stage. In the sporophytic system the stigma is papillate and dry and covered with hydrated layer of protein known as pellicle. This pellicle is involved in incompatibility reaction. With in few minutes of reaching the stigmatic surface the pollen releases an exineexcaudate which is either protein or glycerol protein. This reacts with pellicel and induces cellulose formation which further prevents the growth of pollen tube.

2. Pollen Tube-style interaction

There is germination of pollen grains and pollen tube penetrates the stigmatic surface. But due to incompatible combinations the growth of pollen tube is retarded with in the style. Style interaction is also found in *Petunia, Lycopersicon* and *Lilium*. The protein and poly saccharide synthesis in the pollen tube is prevented resulting in bursting up of pollen tube and resulting death of nuclei.

3. Pollen Tube-ovule interaction

In some species, the SI reaction occurs only when the pollen tube reach the ovary and fertilization occurs but the embryo degenerates later due to some biochemical reaction. Such interaction has been reported in *Theobroma cacao.*

Overcome of Self-Incompatibility

1. Bud pollination

Pollination of immature bud with mature pollens is find useful. The bud pollination should be done 2-4 days before opening of a flower. This method is now used for production of inbred lines for hybrid seed production in Brassica.

2. Delayed pollination

Delayed pollination leads to self fertilization in some self incompatible species. This has been reported in Brassica and Lilium that pollination of aged pistil several days after maturity with normal incompatible pollens resulted in some degree of self fertilization.

3. Late season pollination

Self pollination at the end of flowering season leads to seed set. The case has been reported in Nicotiana, Petunia and *Abutilon*.

4. Irradiation

When style is irradicated with X-rays immediately before selfing resulted in breakdown of SI. Ex., Petunia

5. High temperature

Treatment of style at temperature ranging from 30 to 60°C leads to break down of SI. Significant results have been reported in Malus, Pyrus, Prunus, Trifolium, Lycopersicon and Rye.

6. *In vitro fertilization*

Placing of pollen grains in direct contact with ovules resulted break down of SI in many crop species.

Importance of Self-Incompatibility in Plant Breeding

1. In self incompatible fruit trees, plant two cross-compatible lines to ensure fruitfulness.
2. The use of self-incompatibility in hybrid seed production, two self-incompatible but cross compatible lines are to be inter planted, seeds harvested from both the lines would be hybrid seed.
3. Alternatively, a self incompatible line may be inter planted with a self-compatible line, seed from only the self incompatible line would be hybrid seed.

4. This system does not required emasculation.
5. Does not require resorting to genetic or cytoplasmic male sterility.
6. This system permits combining of desirable genes in a single genotype from two or more different sources through natural cross pollination which is not possible in self-compatible species.
7. SI has been utilized for production of commercial hybrids in Brassica (cabbage and Brussels sprouts) and sunflower.
8. Moreover, knowledge of SI especially in fruit crops helps fruit growers to increase the yield of fruits by providing suitable pollinators.
9. In case of pineapple, commercial clones are self incompatible. As a result, their fruits develop parthenocarpically and are seedless.

Limitation of Self-Incompatibility

1. Difficult to produce homozygous inbred lines in a self incompatible species.
2. For maintenance of the parental lines, bud pollination has to be made .
3. SI is affected by environmental factors like, temperature and humidity.
4. SI is broken down at high temperature and humidity.
5. Sometimes sib mating have been reported in case of Brussels, when bees visit only one parental line.

7

Pollination in Plants

Pollination can be defined as the pre-fertilization event or process where pollen grains from anther are transferred to the stigma of a flower. Pollination is the process that helps to unite the male and female gametes and thus helps in fertilization. It can be broadly classified into two, cross-pollination and self-pollination.

1. Self-Pollination

Self-pollination is the process by transferring the pollen grains directly from anther into the stigma of the same flower. It is called as self-pollination. This form of pollination is common in hermaphrodite or dioecious plants which have both male and female sexual organs on the same flower. In self-pollinating flowers, the anthers and stigma are of similar lengths to facilitate the transfer of pollen. Self – pollination can be further divided into two types:

- ***Autogamy***: In this type of self-pollination, the pollen is transferred from the anthers of one flower to the stigma of the same flower.
- ***Geitonogamy***: In this type of self- pollination, the anthers are transferred from the anthers of one flower to the stigma of another flower but on the same plant.

Mechanism to Promote Self-Pollination

Various mechanism which promote self pollination, are briefly given below :

1. **Cleistogamy :** Pollination occurs before opening of a flower or pollination and fertilization takes place in an unopened flower bud. Example., wheat, barley, oat etc.
2. **Chasmogamy :** Flowers open after completion of pollination. Example., wheat, barley, oat and rice.
3. **Bisexuality :** Presence of both male and female flowers on the same plant.
4. **Homogamy :** When anthers and stigma mature at the same time.
5. **Position of anthers :** When stigma is surrounded by anthers. Example., tomato and brinjal.

Advantages of Self – Pollination

1. They have regular self pollination.
2. They are homozygous and have advantage of homozygosity, means they are true breeding.
3. There is no diversity in the genes and therefore, the purity of the race is maintained.
4. The plants do not depend on external factors for pollination and even smaller quantities of pollen grains produce success rate in getting pollination.
5. Self-pollination ensures that recessive characters are eliminated.

Disadvantages of Self-Pollination

1. When there are no more genes, there are no or few new characters or features that are introduced into the offspring's.
2. It reduces the vigor and vitality of the race in the absence of new features.
3. Without new characters introduced, the resultant offspring's' immunity to diseases reduces.

2. Cross-Pollination

Cross pollination during which the pollen grains are transferred from the anther of one flower into the stigma of another flower. In this case, the two flowers are genetically different from each other. Cross-pollination is always dependant on another agent to cause the transfer of pollen.

Mechanism to Promote Cross-Pollination

There are several mechanism that promotes cross pollination which are summarized below:

1. **Dioecy** : When male and female flowers are born on different plants. Example., papaya, datepalm, hemp, spanish and asparagus.
2. **Monoecy** : When both male and female flowers are born on the same plant, either on the same inflorescence, eg., banana, coconut, castor, mango or on the separate inflorescence, cucurbits, strawberries, grapes etc.
3. **Protandry** : When anthers mature before stigma. Example., maize, sugar beet.
4. **Protogyny** : When stigma matures before pollen grains. Example., pearl millet, isabgol.
5. **Male sterility** : Non functional pollen grains. Used in hybrid seed production.
6. **Self incompatibility** : The inability of the functional pollens to fertilize the same flowers. Example., brassica, radish, nicotiana etc.

Advantages of Cross-Pollination

They have random mating. In such population, each genotype has equal chance to mate with other genotypes.

1. Individuals are heterozygous and have advantage of heterozygosity.
2. Cross-pollination is beneficial to the race of the plant as it introduces new genes into the lineage as a result of the fertilization between genetically different gametes.
3. Cross-pollination improves the resistance of offsprings to diseases .
4. The seeds produced as a result of cross-pollination are good in vigor and vitality.
5. If there are any recessive characters in the lineage, they are eliminated as a result of genetic recombination.
6. It is the only way that unisexual plants can reproduce.

Disadvantages of Cross-Pollination

1. There is a high wastage of pollen grains that need to be produced to ensure fertilization occurs.
2. There are high chances that the good qualities may get eliminated and unwanted characteristics may get added due to recombination of the genes.

The crop plants belongs to self and cross pollination are presented in Table 7.1.

Table 7.1. Classification of crop plants based on mode of pollination and reproduction

Mode of pollination and reproduction	Crop plants
Self Pollinated Crops	Rice, Wheat, Barley, Oats, Chickpea, Pea, Cowpea, Lentil, Green gram, Black gram, Soybean, Common bean, Moth bean, Cluster bean, Linseed, Sesame, Horsegram, Sunhemp, Chillies, Brinjal, Tomato, Okra, Peanut, Potato, *etc.*
Cross Pollinated Crops	Mize, Pearl millet, Rye, Sunflower, Safflower, Castor, Alfalfa, Radish, Cabbage, Sugarbeet, Red clover, White clover, Spinach, Onion, Garlic, Turnip, Squash, Muskmelon, Watermelon, Cucumber, Pumpkin, Oilpalm, Date palm, Carrot, Coconut, Papaya, Sugarcane, Coffee, Cocoa, Tea, Apple, Pears, Peaches, Cherries, grapes, Almond Strawberries, Pine apple, Banana, Cashew, Irish, Cassava, Taro, Rubber, *etc.*
Often Cross Pollinated Crops	Sorghum, Cotton, Triticale, Pigeonpea, Tobacco.

8

Hybridization Techniques

The mating or crossing between two genetically dissimilar plants or lines is known as hybridization. In other words, hybridization is an artificial controlled pollination in which pollen grains of desired male parent is transferred on the stigma of the desired female parent.

Objectives of Hybridization

1. To create genetic variability, which is pre-requisite for crop improvement programme.
2. To transfer one or more characters into single plant.
3. To know the inheritance of the characters.
4. To estimate the general combining ability of the parents and specific combining ability of the crosses.
5. To exploit and utilize the hybrid vigour.
6. To produce hybrid/synthetic varieties.

Types of Hybridization

1. Inter varietal/intra specific hybridization

The parents involved in hybridization belong to same species or they may be two strains or varieties or races of the same species. This type of hybridization is most commonly used in crop improvement programme.

2. Inter-specific/ intra generic hybridization

Crosses are made between the plants of two different species of the same genus. In other word, the crossing between two species of the same genus. Example: *T. aestvum x T.durum, Oryza sativa x O. perennis*.

3. Inter-generic hybridization

When a cross is made between the plants belonging to different genera. It is used to transfer desirable characteristic from one genus to another genus. Example: *Triticum spp x Scale cereal.*

4. Introgressive hybridization

In this type of hybridization hybrids may be repeatedly backcross to one of the parental species. The most of the genotype of that parental species would be recovered along with the genes from other parental species. Example: primitive maize x Tripsacum produces modern maize.

Basic Requirements for Hybridization

1. Carefully study the structure of flower before starting the operations.
2. Select those flowers which produce more and better seed.
3. Learn timing of opening of the flower.
4. Know the timing that pistil remain receptive and pollen grains are capable of functioning.
5. Procure necessary instruments.
6. Generally crosses are attempted in the evening.
7. Normally pollination is done in the morning.

Steps of Hybridization

1. Selection of parents

Selection of parents is first and most important step of hybridization. At least one parent should be of well adapted variety. Parents should posses the desirable characters.

2. Selection of flowers

Select immature and healthy flowers prior to anthesis. Retain only few flowers in which crossing is done and remove all other flowers

3. Preparation of female flower

Transfer of pollens from desired male parent to the stigma of the emasculated flower of female parent is known as pollination. Select the plants in which emasculation is carried out one or two days prior to emasculation. Pollen parent is also selected from pollens are used in pollination. Pollination is generally carried out in the morning, when stigma is receptive. Fresh and viable pollens are collected from the male parent & are dusted on the emasculated flower of female parent.

4. Bagging/ Labeling

Just after pollination, the emasculated flowers are bagged with the crossing bags and labeled.

5. Harvesting of cross seed

At maturity, bags are removed and cross seeds are harvested separately and collected along with the label.

Emasculation

Removal of stamens or anthers or killing the pollens of a flower without damaging the female reproductive organ before pollination is known as emasculation. In bisexual flowers, emasculation is essential to prevent the self-pollination. In monoecious plants, male flowers are removed in castor and coconut or male inflorescence is removed in maize. In species with large flowers e.g. (cotton and pulses) hand emasculation is accurate and adequate.

Methods of Emasculation

1. Hand emasculation

In hand emasculation, removal of anthers from the flower with the help of forceps without damaging the stigma. It is done before anther dehiscence. Emasculation is generally done between 4 to 6 PM one day before anthers dehisce. It is always desirable to remove other young flowers located close to the emasculated flower to avoid confusion. The corolla of the selected flower is opened and the anthers are carefully removed with the help of forceps. Sometimes corolla may be totally removed along with epipetalous stamens. In cereals, one third of the empty glumes will be clipped off with scissors to expose anthers. In wheat and oats, the florets are retained after removing the anthers without damaging the spikelet. In all the cases, gynoecium should not be injured. An efficient emasculation technique should prevent self pollination and produce high percentage of seed set on crosspollination.

2. Suction method

This method of emasculation is useful in the species with small flowers like *cotton*. Emasculation is done in the morning immediately after the flowers open. A thin rubber or a glass tube attached to a suction hose is used to suck the anthers from the flowers. The amount of suction used is very important which should be sufficient to suck the pollen and anthers but not gynoecium. In this method considerable self-pollination, upto10% is like to occur. Washing the stigma with a jet of water may help in reducing self-pollination however self pollination cannot be eliminated in this method.

3. Hot water treatment

Pollen grains are more sensitive than female reproductive organs to both genetic and environmental factors. In case of hot water emasculation, the temperature of water and duration of treatment may vary from crop to crop. It is determined for every species. For sorghum 42-48°C for 10 minutes is found to be suitable. In the case of rice, 10 minutes treatment with 40-44°C is adequate. Treatment is given before the anthers dehiscence and prior to the opening of the flower. Hot water is generally carried in thermos flask and whole inflorescence is immersed in hot water.

4. Alcohol treatment

In this method, inflorescence is immersed in the alcohol of suitable concentration for a short period followed by rinsing with water. In Lucerne the inflorescence immersed in 57 % alcohol for 10 second was highly effective. It is not commonly used.

5. Cold treatment

Cold treatment like hot water treatment kills the pollen grains without damaging gynoecium. In the case, rice treatment with cold water 0.60°C kills the pollen grains without affecting the gynoecium. This is less effective than hot water treatment.

6. Genetic emasculation

Genetic/cytoplasmic male sterility may be used to eliminate the process of emasculation. This is useful in the commercial production of hybrids in maize, sorghum, pearl millet, onion, cotton, rice etc. In many species of self-incompatible cases, also emasculation is not necessary, because self-fertilization will not take place.

7. Use of gametocide

It is also known as chemical hybridizing agents (CHA). There are some chemicals, which selectively kills the male gamete without affecting the female gamete. Example: Ethrel, Sodium methyl arsenate, Zinc methyl arsenate in rice, Maleic hydrazide in cotton and wheat.

9

Breeding Methods for Self Pollinated Crops

The prime object of plant breeder is to develop high yielding variety with superior characteristics. Selection is based on the phenotypes of the plants. Consequently the effectiveness of selection primarily depends on the degree to which the phenotype of plants reflects their genotype. These are all self fertilizing species. In these species development of seed take place by self fertilization. Hence self pollinated species are also known as autogamous species. Various plant characters such as homogamy, cleistogamy, chasmogamy, bisexuality etc. promotes self pollination. The self pollinated crops with their description are given in Table 9.1. The breeding methods used in self and often-self pollinated crops are discussed below:

Plant Introduction

Plant introduction consists of taking a genotype or a group of genotypes of plants into a new area or region or environment, where they have not being grown earlier. It is very simplest and fastest method of developing new variety. It is of two types:

1. Primary introduction

Introduction refers as the introduced material that can be used for commercial cultivation as a variety without any change in the original genotype. For example- varieties sonara-64 and Lerma rojo in wheat.

2. Secondary introduction

The material that can be used as a variety after selection from the original genotype or may be used for transfer of some desirable gene to the cultivated variety. For example- wheat variety of Kalyan Sona and Sonalika.

Plant Introduction Agencies in India

1. **NBPGR** (National Bureau of Plant Genetic Resources), IARI, New Delhi

NBPGR has ten regional stations:

i) Akola (Maharashtra)
ii) Bhowali (Uttar-Pradesh)
iii) Hyderabad (Andhra-Pradesh)
iv) Jodhpur (Rajasthan)
v) Shillong (Meghalaya)
vi) Shimla (Himachal Pradesh)
vii) Thrissur (Kerala)
viii) Srinagar (J&K)
ix) Cuttack
x) Ranchi

2. **Forest Research Institute**, Dehradun
3. **The Botanical Survey of India**, Kolkata

Pure Line Selection

Pure-line is the progeny of a single self-fertilized homozygous crop plant. The term pure line and pure line method was suggested by Johannsen (1903) while working with the "princess variety of French beans *(Phaseolus vulgarizes*). Pure line varieties are homozygous and homogeneous. All the plants within a pure-line has identical genotype and if there is any variation within a pure-line is solely due to the environment. In this selection method, numerous superior looking plants are selected from the genetically variable population and selected plants are harvested separately and individual plant progenies are evaluated by simple observation, frequently over a period of several years. Pure line is used for improvement of local or desi varieties, which have cultivated for long time. Therefore, pure-line selection is also known as individual plant selection.

Procedure

First year: 500-2000 superior plants are selected on the basis of phenotypes from desi or old variety and selected plants are harvested individually.

Second year: Individual plant progenies are grown, superior progenies are selected. Undesirable and suspected progenies are rejected.

Third Year: Selected progenies are planted in preliminary yield trial along with standard check.

Fourth to sixth year: Multi location yield trials along with standard checks are conducted at several locations to see their performance. Disease resistance and quality tests are also done.

Seventh year: Highest yielding genotype is put forwarded as a new cultivar. Progeny showing superiority to check variety is released as a new variety. Seed multiplied for distribution for cultivation.

Advantages of Pure line Selection

1. Pure line reflects maximum genetic advance from a variable population.
2. All plants are of similar types.
3. Selection based on progeny performance is effective for characters with relatively low heritability.
4. Used for improvement of desi varieties.

Disadvantage of Pure line Selection

1. Requires more time than mass selection to develop new cultivar.
2. Less adaptable to fluctuating environments.
3. Leads to fast genetic erosion.
4. Does not create new variations.

Achievements

- **Rice** : Mtu-1, Mtu-3, Mtu-7, Bcp-1, Adt-1, 3, 5 and 10, GEB-24, BASMATI-370, SAFRI-17.
- **Sorghum** : G 1 and 2, M 1 and 2, CO 1, 4 and 5.
- **Groundnut** : TMV 3, 4, 7, 8 and Kadiri 71-1.
- **Redgram** : TM-1, ST-1, Malviya Vikalp.
- **Chilies** : G1 and G2.
- **Ragi** : AKP 1 to 7.
- **Wheat** : NP4, NP12, NP6, PB8, PB9D, K13, K46, K54.
- **Barley** : RS17 (1st variety released in Rajasthan).
- **Soyabean** : JS71, JS2, JS93-05, MAUS2, NRC7, Shilajeet, VLS2, VLS21, CO1, ADT1.
- **Mung-bean** : Kopergaon, CO1, CO2, CO3, Gujarat, shining mung, T1, B1.
- **Lentil** : L9-12, TYPE8, TYPE36, BR75, C31, PANT L406.
- **Cotton** : MCU5VT, COCANADA2, GAURANI22, GAURANI46, LOHIT, DHARWAD AMERICAN

Mass Selection

Mass selection is the oldest and easiest type of selection. It has been practiced from the time of domestication It is suggested by Gardner in 1961. In this method a large number of phenotypically superior plants are selected from the base population, harvesting and bulking their seeds together to constitute a new variety. This process is repeated till desired characters are obtained. Therefore, selection is done for easily observable characters. Such as plant height, number of tillers, ear type, ear length, disease resistance, lodging resistance, shattering resistance etc.

Mass selection can be practiced both in self and cross pollinated crops. This method is used for the improvement of old varieties. In the local or old variety, several plants would be inferior and low yielding, such plants will be eliminated through mass selection and local variety would be improved without adversely affecting its adaptability and stability.

Procedure

First year: Large number of phenotypically desirable plants are selected. The selected plants are harvested and their seeds mixed together to grow next generation.

Second year: The bulk seed of selected plants is planted in preliminary yield trial along with check variety.

Third to fifth year: The performance of genotypes is evaluated in multi-location yield trials.

Sixth year: Best genotype is identified and released as a variety for commercially cultivation.

Advantages of Mass Selection

1. To increase the frequency of desirable genotypes from a genetically variable population.
2. Purify a mixed population with differing phenotypes.
3. Adaptation of the original is not change.
4. Improving the average performance a new cultivar.
5. Retain considerable genetic variability.

Disadvantages of Mass Selection

1. Mass selection is not effective with low heritability traits.
2. Without progeny testing, heterozygote's can be inadvertently selected.

3. Mass selection cannot create new variability.
4. Final product is not as uniform as those developed through pure-line selection.

Achievements

- **Groundnut** : TMV-1 and TMV-2.
- **Bajra** : Pusa moti, Baba puri, Jamnagar gaint, AF3.
- **Sorghum** : R.S. 1.
- **Rice** : SLO 13, MTU-15.
- **Potato** : K122.
- **Cotton** : Bikanerinerma, F414, H777, Pramukh, 8RT, PRS72, Mysore Vijay.

Pedigree Method

The pedigree is refer as the description of the ancestors of an individual and it is generally helpful in finding out the amount of relationship among two individuals, i.e., whether they are related by common parent in their descent ancestry. Pedigree method is used for genetic improvement of self-pollinated crop species in which superior genotypes are selected from segregating generations and proper pedigree records of selected plants are maintained in each generation. So that it is possible to trace pedigree line to a specific plant of F_2 generation. In pedigree method of breeding, individual plants are selected from F_2 and subsequent generations and their progenies are evaluated. This method is most suitable for improving specific traits like, disease resistant, days to maturity, plant height etc. Moreover, it is more commonly used for the improvement of polygenic traits. In cross pollinated species it is used for development of inbred lines.

Applications of Pedigree Method

1. This method is widely used for improvement of self pollinated species for development of new pure line varieties.
2. This method is also used in cross pollinated species for development of inbred lines.
3. It is commonly used for improvement of polygeneic traits than oligogenic traits.
4. Generally used when both the parents, which are used in hybridization programme have good agronomic characters.
5. This method is also suitable in selection of new recombinant types from segregating population.
6. It is almost hoped that some transgressive segregants would be recovered.

Breeding procedure

First year: First parents are selected as per the breeding objective and planted in crossing blocks. The crosses are attempted between selected parents. The F_1 seeds are harvested separately from each plant.

Second year: F_1 generation materials are planted with wider spacing and selfing is allowed, each F_1 will produce more F_2 seeds. The F_2 seeds are harvested.

Third year: Planting large F_2 population (2000-10000 plants) and selection of a large number of plants (200-500). Individual plant selection is practiced in F_2. The progeny of each selected plant is harvested separately.

Fourth year: Each plant progeny is grown separately. Individual plant progenies with desirable characteristics are selected and inferior progenies are rejected.

Fifth year: The selection procedure is repeated as previous year. Some progenies may appear fairly uniform, but several may still segregate. A few best plants may be selected from each of the best progenies.

Sixth year: Individual plant progenies of F_5 generation are planted according to recommended commercial seed rate. Most of the progenies may reach to the homozygosity. If the progenies show variation then the individual plants are selected. Generally, segregation does not continue beyond the F_6 generation.

Seventh year: Preliminary yield trial may be done for the progenies which are reasonably homozygous and have enough seeds.

Eighth year: Preliminary yield trial with three or more replications is conducted to identify few superior lines. The progenies are evaluated for flowering, maturity, plant height, disease resistance, no. of effective tillers, primary and secondary branches, yield and quality test etc.

Ninth to tenth or thirteenth year: The promising genotypes are tested in replicated yield trials along with standard checks at several locations. The genotype which is found superior to the best commercial variety, may be identified.

Eleventh or fourteenth year: The identified genotype should get multiplied to release as a new variety. Thus release of new variety by takes about 14-15 years. Then certified seed is multiplied for cultivation. The procedure is outlined in Figure 9.1.

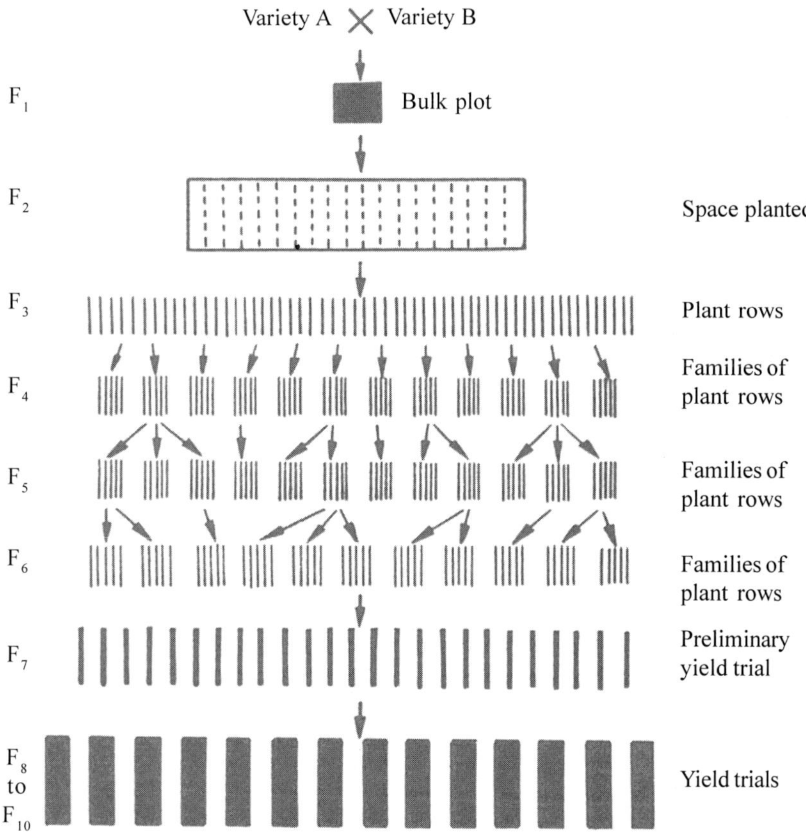

Fig. 9.1. Systematic representation of pedigree method

Advantages of Pedigree Method

1. The pedigree method permits the breeder to use his skill in selection of plants to a greater degree than any of the other methods.
2. Selection in each generation involves a different environment (year) which provides a good opportunity for expression and selection of important characters.
3. The genetic relationships of lines are known. This can be utilized to select multiple lines from specific desirable families or to maximize genetic variability among lines retained during selection by avoiding closely related individuals whose probable worth may be identical.
4. This method provides the chances of recovering transgressive segregats.
5. The selection is based on genotypic value.

Disadvantages of Pedigree Method

1. It is very difficult to handle large population.
2. Required much more time for keeping the record of selected plants.
3. Breeder has to devote more time.
4. Requires more land and labor than other methods of inbreeding.

Achievements

- **Wheat**- NP-52, 120,125, 700 and 800 series, K65, K68, HD 2281, HD 2285, HD 2380, ND 2402, Janak, RAJ 2535, DWR 39, WH331, WL 616.
- **Rice-** ADT-25, Jaya, Padma, Bala, Kauveri, Karuna, Krishana, Ratna, Sabarmati.
- **Cotton-** J 34, J 205, LH 372, LH 900, LH 1556, F 846, F 1054, RST 9, Vikas, H 1089, Sharda, LRA 5166, Anjali, Surabbi, Suvin, HS 6, H 655C, SH 131, MCU 8, MCU 9, Lakshmi, Digvijay,
- **Sorghum**- Co 18, Rs 610, J34, S205, LH372, Khandwa 3, Rajat, Nagnath, Srisaitam, Mahanandi, Jyoti
- **Chickpea**- T 1, T 2, T 3, T 5, Radhey.
- **Greengram**- T 2, T 44, T51, Sheela etc.
- **Pigeonpea**- T 21, Prabhat
- **Pea-** Pant Matar 2, Jawahar Matar 1, Jawahar Matar 4.

Bulk Breeding Method

In bulk method, selection is not practiced till F_2 to F_6. The plants are harvested as a bulk to raise the next generation. This practiced remain continued upto F_5 or F_6 generation. By the end of F_6 generation a high proportion of plants attained the homozygosity. Bulk plot of F_2 to F_6 may be subjected to disease, pests and other stresses to eliminate the susceptible genotypes. In F_7 generation large number of plants is space planted, superior plants based on phenotypes are selected and their seeds are harvested separately. This is continued upto F_9 when promising lines are selected and advanced to preliminary yield trial. In F_{10} to F_{12} replicated yield trials along with commercial varieties as checks are conducted over several locations (Figure 9.2). Best one is released as a variety. In this method predominance of selection is natural selection. The bulk method is simple, convenient, inexpensive and less labour consuming, but time consuming. Bulk method was first proposed by Nilsson Ehle in 1908 at Svalof. This method is also known as mass method 'or' population method of breeding.

Features of Bulk Method

1. This method is suitable for isolation of homozygous lines.
2. Waiting for occurrence of suitable environments for selection of suitable plants.
3. Providing an opportunity for natural selection to change the composition of population.

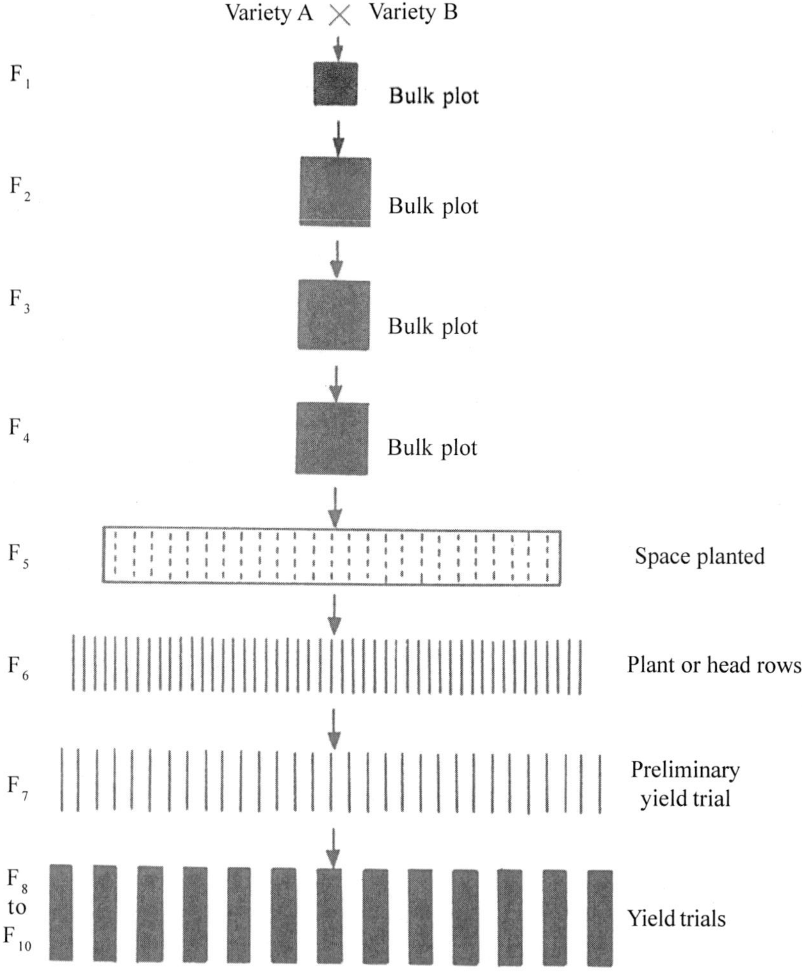

Fig. 9.2. Generalized scheme of bulk method

Advantages of Bulk Method

1. Simple, convenient and inexpensive
2. No pedigree record is to be kept, which save time and labour.
3. Little work and attention is needed for handling the segregating populations
4. Natural selection plays an important role in elimination poor genotypes.
5. There is more chance for isolation of transgressive segregates than in pedigree method.
6. Selection for some environment dependent characteristics like disease resistance, cold and high temperature tolerance can be carried out when environment is favorable for disease epidemic.
7. Individual plant selection is carried out, when population reach to homozygosity.

Disadvantages of Bulk Method

1. It takes much longer time to develop a new variety.
2. Informations of inheritance of characters cannot be obtained due to non maintenance of pedigree record.
3. This method is not suitable for greenhouses.
4. Large number of progenies has to be handled at the end of bulking period.
5. Natural selection may also work against desirable traits.
6. It provides little opportunity for a breeder to use his skills.
7. No recombination occurrence among superior lines

Achievements

The method has been used to a limited extent is Barley breeding in U.S.A. and more than 50 varieties were developed. They are Arival, Beecher, Glacier, and Gem originated from a cross Atlas x Vaughn. In India only one variety Narendra Rai in brown mustard has been developed by bulk method.

Single Seed Descent (SSD) Method

Single seed descent method is a modification of bulk method of breeding. This method was first suggested by Goulden in 1941. In this method parents are selected as per the breeding objectives, both the parents are crossed to get F_1. F_1s are space planted to get more F_2 seeds. Single seed from each plant is taken in F_2 generation and seed from all the plants are bulked to raise the F_3 generation. Same procedure is repeated to grow F_4 and F_5 generation when the plants reached to nearly homozygous (Figure 9.3). In F_6 generation large number of individual plants are selected to raise the next generation. This system is repeated upto F_8. After that preliminary and coordinated yield trials are conducted. Best one is released as variety. This method is specially suited for advancing generation of rice crosses. This is especially suited to the characters with low heritability.

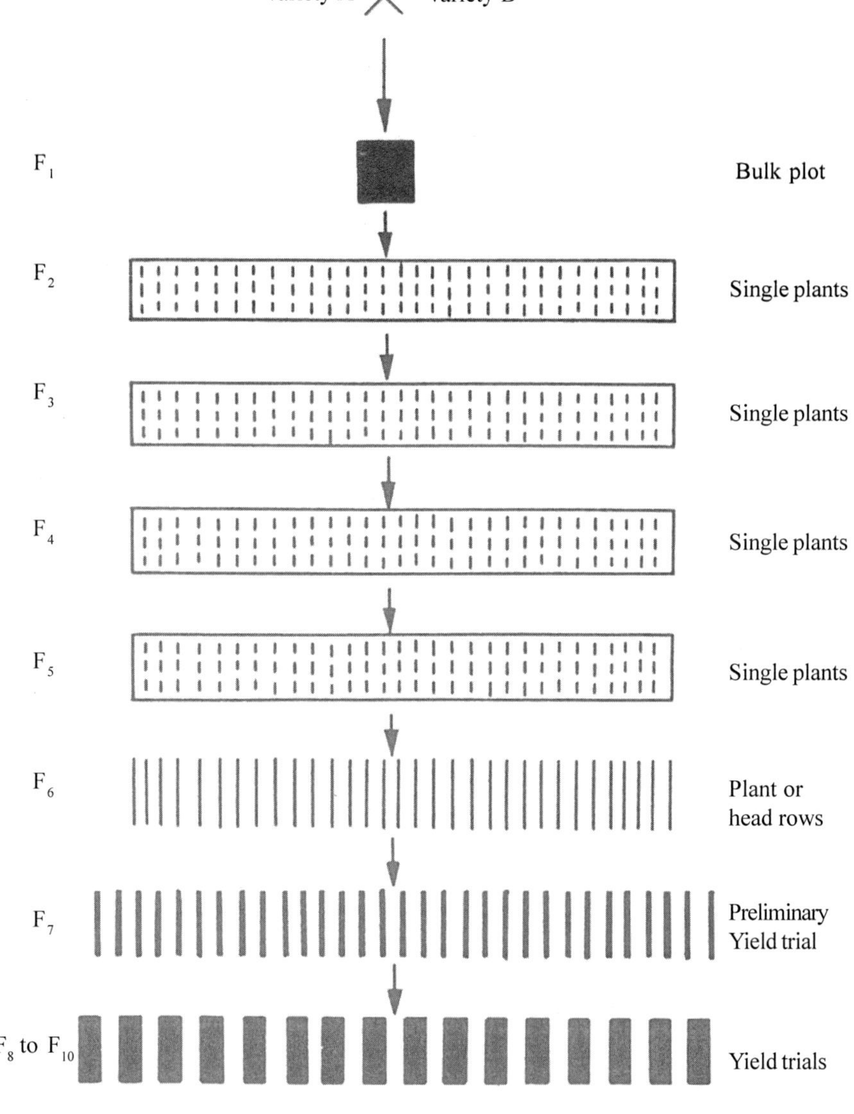

Fig. 9.3. Systematic representation of SSD method

Schematic representation of single descent method

Selected plants are hybridized.

F_1 space planted harvested in bulk.

F_2 densely planted. From each plant one random seed selected and bulked.

F_3 as in $F_{2.}$ F_4 as in F_2.F_5 as in F_2.

F_6 space planted. 100-500 plants with desirable traits are harvested separately.

F_7 individual plant progenies grown.

Undesirable progenies eliminated. Desirable homozygous progenies harvested in bulk.

F_8 individual plants may be selected in outstanding progenies showing segregation.

F_9-F_{10} PYT with suitable check.

Coordinated yield trials, diseases and quality tests.

F_{11}-F_{14} seed increase for distribution.

Advantages of SSD Method

1. Advances the generation with the maximum possible speed.
2. Requires less space, time and resources in early stages.
3. Superior to bulk/mass selection.
4. Delayed selection eliminated confusing effects of heterozygosity; more effective than pedigree breeding when dealing with low heritability traits;
5. Retain considerable variability.
6. Variation for quantitative traits with low heritability is maintained using SSD.

Disadvantages of SSD Method

1. It does not permit any form of selection during the segregating generations.
2. Fewer field evaluations, so lose the advantage of natural selection.
3. Need appropriate facilities to allow controlled environment manipulation of plants for rapid seed production cycles.
4. Natural selection usually has little impact on population.
5. Identity of superior F_2 plants is lost This prevents the breeder from examination superior F_2 families.

Achievements

- Pea: Ultara

Backcross Method

Backcross method was proposed by Harlan and Pope in 1922. A cross between a F_1 hybrid and one of its parents is known as backcross. In back cross method the hybrid and the progenies in the subsequent generation are repeatedly backcrossed to one of the parents of the F_1. The backcross breeding approach can be employed to introduce a specific trait, such as disease resistance, from one genotype to another genotype. The objective of backcross method is to improve one or two specific defect of a high yielding and well adapted variety. In backcross method, the parent from which few or more genes are transferred to the recurrent parent is termed as donor parent or non-recurrent parent and the parent to which genes are transferred from the donor parent is called the recurrent parent or recipient parent. Donor parent is used singly, whereas recurrent parent is used repeatedly in this method. Backcrossing involves making an initial cross between the donor and recurrent parents. The resultant F_1 progeny have 50% of their genetic material from each parent. F_1 individuals are crossed to the recurrent parent to develop a backcross one (BC_1) population. Individuals from the BC_1 population are once again crossed to the recurrent parent. Each generation of backcrossing reduces the proportion of the donor parent present in the population by half. This cycle of crossing backcross progeny to the recurrent parent continues until a new line that is identical to the recurrent parent, but with the desired gene or trait from the donor parent is created. By the BC_4 generation, the lines are 96% more than identical to the recurrent parent.

Application of Backcross Method

1. Inter- varietal transfer of simply inherited characters.
2. Inter-varietal transfer of quantitative characters.
3. Transfer of cytoplasm.
4. Transgressive segregants.
5. Production of isogenic lines.
6. Germplasm conservation.

Genetic Consequences

1. Repeated backcrossing increases homozygosity at the same rate as selfing.
2. The genotype of backcross progeny becomes similar to that of recurrent parent.
3. There would be opportunity in each back cross generation for crossing over to occurs between the genes being transferred and genes tightly linked with it.
4. When genes are tightly linked, large back cross population would have to be grown to obtain desirable recombinant types.

Requirement for a Backcross Programme

1. A desirable recurrent parent, which lacks in one or two traits. 2. A suitable donor parent that has gene for disease resistance. 3. The character to be transferred from donor to recurrent parent must have high heritability and governed by one or few genes. 4. Sufficient number of backcrosses, 4-6 should be made, so that genotype of the recurrent parent is recovered in full.

Procedure of backcross method for the transfer of dominant gene

- **Hybridization** : A cross is made between variety A, having rust resistance gene, used as a donor parent with variety B which is rust susceptible but agrinomically desirable.
- **F_1 generation :** Selection for disease resistance is not performed. Plants from F_1 seeds are back crossed with recipient variety B and seeds are collected to raise BC_1 generation.
- **First backcross generation (BC_1)** : Disease resistance plants are selected and back crossed with recurrent parent B.
- **Second to fifth backcross generation (BC_2 to BC_5)** : Segregation for disease resistance occurs in every back cross generation. Plants are selected on the basis of similarity with recurrent parent and resistance to disease. Selected ones are back crossed with recurrent parent B.
- **Sixth backcross generation (BC_6)** : Rust resistant plants are selected, self pollinated and harvested separately.
- **BC_6 F_2 generation** : Individual plant progenies are grown. Plants are selected on the basis of similarity with recurrent parent and at the same time resistance to disease. They are harvested separately.
- **BC_6 F_3 generation** : Individual plant progenies are grown and progenies similar to B are harvested and bulked.
- **Yield trials** : Replicated yield trials are conducted with recurrent parent as a check. The newly variety should be similar to variety B for most of the important characteristics with resistant to disease. Variety is identified for released and seeds are multiplied for distribution to farmers.

Procedure of backcross method for the transfer of recessive gene

- **Hybridization** : The recurrent parent (A) is crossed with rust resistant donor parent (B). The recurrent parent is generally used as female. i.e (RR X rr).
- **F_1 Generation**: No rust resistance test and F_1 plants are backcrossed to the recurrent parent.

- **BC_1 Generation :** If rust resistance is recessive all the plants will be rust susceptible. The plants are not selected for rust resistance. All the plants are self- pollinated.
- **BC_1F_2:** Rust resistant test is done and resistance plants are selected and back crossed to recurrent parent.
- **BC_2 :** Rust resistance is not carried out and plants back crossed to A.
- **BC_3 :** Rust resistance is not carried out and all plants are self pollinated.
- **BC_3 F_2 Generation :** Plants are inoculated with stem rust. Rust resistant plants, similar to 'A' are selected and backcrossed to variety 'A'
- **BC_4 Generation :** No rust resistance test , plants are backcrossed to variety 'A'.
- **BC_5 Generation :** No rust resistance test, plants are self pollinated.
- **BC_5 F_2: Generation**: Plants are subjected to rust epidemic, resistance plant for rust and having similar characteristic of variety. 'A' is selected and self seeds are harvested separately.
- **BC_5 F_3:** Individual plant progenies are grown and progenies resistance to rust and similar to A are selected and composited.
- **Yield Test**: The new variety is tested in replicated yield trials along with the variety 'A' as a check. The newly constitute variety should be similar to the variety A for most of the important characters. Seeds are multiplied for distribution.

Advantages of Backcross Method

1. It is a conservative method, not permitting new recombination to occur.
2. It is useful for introgression specific genes from wide crosses.
3. The new variety is nearly identical with that of recurrent parent.
4. The backcross method is not dependent upon the environment.
5. It is applicable to breeding both self-pollinated and cross-pollinated species.
6. Back cross method is not changed the adaptability of varieties but replaces the undesirable allele at particular locus
7. This method is used for the development of isogenic lines.
8. Extensive tests are not required.
9. Male sterility and fertility restoration genes can be transferred to various back ground.
10. The variety developed by back cross is easily accepted by farmers.

Disadvantages of Backcross Method

1. Newly developed variety cannot be better than recurrent parent except for the character transferred.
2. It requires 6-8 back crosses, is often difficult and time consuming.
3. Sometime undesirable gene linked may be transferred with desirable one.
4. Backcrossing is not effective for transferring quantitative traits.
5. Recessive traits are more time consuming to transfer.

Achievements

- **Cotton** : Varieties 170-CO-2 and 134 – CO2M, Vijay, Kalyan, Digvijay.
- Wheat variety Kalyana Sona, which became susceptible to leaf rust. Resistant has been transferred to Kalyan Sona from several diverse sources i.e., Robin, K1, Blue bird, Tobari, Frecor and HS-19. Baart 38 is another variety of wheat developed by Back cross method.
- Tift 23A is susceptible to downy mildew. The line backcrossed with MS-521A, MS-541 A, MS-570A resistant hybrids were produced.
- Three multiline varieties, such as KSML 3, MLKS 11 and KML 7406, have been developed for cultivation.
- Varieties Kh 65, NP 852, NI 5439, IWP 72, HUW 234, K 8027 have been evolved in wheat through back cross method.

Table 9.1. Self pollinated crops with their descriptions

Crops	Botanical name	Family	Chromosome number (2n)	Origin
Mung	Vigna radiata	Leguminaceae	22	India
Urd	Vigna mungo	Leguminaceae	22	India
Rajma	Phaseolus vulgaris	Leguminaceae	22	Central America
Guar	Cyamopsis tetragonoloba	Leguminaceae	14	India
Moth	Phaseolus aconitifolius	Leguminaceae	22	India
Pigeon pea	Cajanus cajan	Leguminaceae	22	Africa
Cowpea	Vigna anguiculata	Leguminaceae	22	Africa
Soyabean	Glycine max	Leguminaceae	40	China
Cotton	Gossypium spp	Malvaceae	26	India
Till	Sesamum indicum	Pediliaceae	26	India
wheat	Triticum aestivum	Graminae	42	South West Asia
Rice	Oryza sativa	Graminae	24	South East Asia
Barley	Hordeum vulgare	Graminae	14	North West India
Oats	Avena sativa	Graminae	42	Asia minor
Pea	Pisum sativum	Leguminaceae	14	Central Asiatic centre
Groundnut	Arachis hypogaea	Leguminaceae	40	Brazil
Gram	Cicer arietinum	Leguminaceae	16	South West Asia

10

Breeding Methods for Cross-Pollinated Crops

The genetic organization of cross pollinated crops are different from that of self pollinated crops, because of differences in the reproductive system. Open pollinated populations are heterozygous with free inter mating individuals, heterozygous at most of the loci. Heterozygosity is an essential feature of commercial varieties of out breeders. It is maintained during breeding or is restored at the end. Population improvement involves selection, generation after generation, with recombination of selected individuals. Its aim to increase the frequency of favorable alleles in the population. The improved forms may be used for commercial cultivation or as a base germplasm for the derivation of inbred lines. The list of cross pollinated plants is presented in Table 10.1. The factors that promote the cross pollination are dicliny, dichogamy, herkogamy, self-incompatibility and male sterility. The breeding methods which are commonly used in cross-pollinated crops are population improvement, hybrid and synthetic varieties.

Population Improvement

In cross pollinated crops heterozygosity must be restored in the end product of any breeding programme. It is classified into the following types:

1. Mass Selection

Mass selection is the oldest breeding method for improvement of cross pollinated crops. A large number of superior plants are selected on the basis of phenotype from the base population and their seeds are bulk together to constitute a variety Therefore, selection is done for easily observable characters like plant height, ear type, grain colour, grains size, disease resistance, tillering ability, lodging resistance, shattering resistance etc. The selected plants are allowed for open pollination. The selection cycle may be repeated one or more times to increase the frequency of favourable alleles. In cross-pollinated crops, mass selection leads to avoid inbreeding depression, loss in vigour and yield. Further because of the heterozygous nature of the population, several cycles of mass selection may effectively be practiced. Mass selection is used for improvement of old varieties. It is not used for handling of segregating generation. Mass selection is successful for improvement of characters with high heritability, it is less effective for yield improvement.

Procedure

First year : i) Large number of plants are selected on the basis of phenotype. ii) Selected plants are harvested and their seeds are bulked.

Second year : i) Bulk seed from the selected plants grown ii) Selection cycle may be repeated. iii) Selection cycle may be repeated one or more year.

Yield trial: Replicated yield trials are conduced along with check variety. Best genotype is identified for released as a variety.

---

2. Modifications of Mass Selection

1. In maize inferior plants are destasselled and the remaining plants are allowed for open pollination. This modification exercise leads to some control of the pollen source.
2. Collect the pollens from all the selected plants and bulked and these pollens are used to pollinate the selected plants. This ensures full control on the pollen source.
3. Stratified mass selection: This modification was suggested by Gardner in 1961. Stratified mass selection is also known as the 'grid method of mass selection'. In this method the field from which selection of the plant is to be made is divided into number of small plots. e.g., 40-50 plants in each plot. Equal numbers of desirable plants are selected within in the plots, not among the plots. The seed from all the selected plants is bulked to grow the next generation. The variation due to environment, will be much reduced with in the small plots than the whole field. It has been able to increase the yielding ability of an open pollinated variety of maize by about 3% per cycle, for15 generations.

3. Ear to Row Method

The simplest form of progeny selection is ear to row method. This method was extensively used in maize by Hopkins in 1908. Superior plants are selected on the basis superior ears from open pollinated population. Next season part of the seed is sown in progeny row and remnant is stored. Superior progeny rows are identified on the basis of evaluation test. Remnant seed of selected entries is composited to constituted materials for the next selection cycle. Ear to row are in fact half sib families. Ear to row method is a powerful tool for increasing the yielding ability of open pollinated varieties of maize. The yielding ability increased at the rate of 3-8 percent per selection cycle.

Procedure

1st Year: i) Large number of plants are selected on the basis of phenotype from the base population. ii) Selected plants are allowed for open pollination. iii) Seeds from each plant is harvested separately.

2nd Year : i) Small progeny rows are grown from the selected plants. ii) Superior progenies are identified. iii) Several phenotypically superior plants are selected from superior plants. iv) Plants are allowed for open pollination and seed harvested separately. The procedure is repeated for more years till the desirable end product is achieved.

---

4. Recurrent Selection

"Reselection generation after generation with interbreeding of selected plants to provide for genetic recombination." The beginning idea of recurrent selection was independently given by Hayes and Garber in 1919 and East and Jones in 1920. The detailed procedure of recurrent selection was described by Jenkins in 1940 and the term of recurrent selection was coined by Hull in 1945.

The method consists of selection of superior followed by selfing and allowed for inter mating. The resulting inter cross population is again utilized as base population to initiate the second cycle of selection. The cycles of selection and crossing continues till desirable results are obtained. The end product is utilized in the production of hybrid and synthetic varieties. This system also maintains high genetic variability and permits recombination among different lines.

Features of Recurrent Selection

1. It is more commonly used in cross pollinated species than in self-pollinated species.
2. A heterozygous base population is required to start recurrent selection viz; open pollinated variety.
3. Synthetic variety, progeny of inter crosses among selected inbreds, double cross and single cross.
4. The population developed by recurrent selection can be used in three main ways : (a) In producing homozygous inbreds by selfing, (b) In the production of hybrid varieties, and (c) In the production of synthetic varieties.

5. Recurrent selection is used to improve the frequency of desirable alleles for a character in a population. In this method the heterozygosity that is lost due to selfing is recovered by intermating of selected progeny.

Types of Recurrent Selection

1. Simple recurrent selection.
2. Recurrent selection for general combining ability.
3. Recurrent selection for specific combining ability.
4. Reciprocal recurrent selection.

1. Simple Recurrent Selection

This type of recurrent selection does not include a tester, also called phenotypic recurrent selection. Due to absence of tester, it does not measure the combining ability. This method is favourable only for high heritability characters.

Main features of simple recurrent selection

- The tester is not used.
- Does not measure the combining ability.
- The selection is based on phenotype.
- Useful only for those characters which have high heritability.

Procedure

First year: i) Phenotypically superior plants are selected from heterozygous base population and they are selfed. ii) Selfed seed harvested individually, evaluated and superior seeds retained.

Second year: i) Individual plants are grown from the selfed seed. ii) Make all possible crosses among progeny. iii) Harvest the crossed seeds in equal quantity and bulked them. This complete the original cycle of selection.

Third year : i) Grow crop from bulk seed of crosses. ii) Again self pollination in a number of plants. iii) Select superior plants at maturity.

Fourth year : i) Progeny of selected plants are grown from selfed seed and intermating is done like first year. ii) The crossed seed is composited in equal quantity for use in the next cycle of selection.

Recurrent selection is effective in increasing the frequency of desirable genes in the population. It is the most suited for character with high heritability. The mean of selected population shifts in the direction of selection.

Systematic Representation

- Several phenotypically superior plants are selected.
- Selected plants self-pollinated.
- Selfed seeds harvested separately.
- Seeds evaluated and superior seeds retained.
- Planted individual plant progenies.
- All possible intercrosses are made.
- Equal amounts of seed from all intercrosses are composited.
- Repeated one or more year.

2. Recurrent Selection for General Combining Ability (RSGCA)

The recurrent selection for GCA was developed as a results of Jenkins (1940) experiments for GCA. Early testing of inbred lines for general combing ability, proposed by Jenkins in 1935. The progenies are crossed with tester with a broad genetic base. So plants are selected on the basis of superior performance of their plant x tester progenies would have superior GCA. This method is used for genetic improvement of quantitative characters. It is a form of half sib recurrent selection.

Main features of RSGCA

1. Used for genetic improvement of quantitative characters.
2. The selection is made on the test cross performance.
3. Generally open pollinated variety is used as a tester for testing general combining ability.
4. Highly effective with incomplete dominance and less effective with over dominance.
5. Used for the improvement of those characters which are controlled by additive gene action.

Procedure

First year: Superior plants for the character under improvement are selected from the base population. The source population may be open pollinated, synthetics or advanced generation of a hybrid. Selected plant are selfed as well as crossed (as male) to a number of randomly selected plants from a tester having broad genetic base. The selfed seed and test cross seeds from each selected plant is harvested seperately.

Second year: Replicated yield trial of test crossed seed is laid out. The superior progenies having high gca are identified.

Third year: The progenies of selected plants with good gca are grown from their selfed seed. These progenies are intercrossed in all possible combinations, equal amount of seeds from all the inter crosses are composited to obtained next generation. This complete original selection cycle.

Fourth year: Seed obtained from bulking of all the inter crosses are planted. Several plants are selected on the basis of their phenotypes. Each of them is selfed as well as crossed with broad genetic base tester.

Fifth year: Operation of second year is repeated.

Sixth year: Operation of third year is repeated. This completes the first recurrent selection cycle.

Seventh year: The second recurrent selection cycle may be initiated.

Thus selection cycles may be repeated till the desired improvement is achieved.

This method improves the GCA in the direction of selection. It increases the yielding ability of the populations to isolate inbreds with superior GCA.

Systematic Representation of RSGCA

- Select several plants on the basis of phenotype.
- Selected plants self-pollinated.
- Selected plants are test crossed from a tester with a broad genetic base.
- Test-cross progeny and selfed seed of each selected plant is harvested separately.
- Replicated yield trial conducted using test cross progeny.
- Identification of superior progenies.
- Selfed seeds from the plants producing superior testcross progenies planted in a crossing block.
- All possible intercrosses are made and
- Equal amounts of seed from all the intercrosses are composited.
- May be repeated.

---

3. Recurrent Selection for Specific Combining Ability (RSSCA)

Recurrent selection for SCA is used to improve the specific combining ability of a population for a character using homozygous tester also called half sib recurrent selection. It is used for genetic improvement of polygenic characters. Selection is made on the basis of test cross performance. This method was originally proposed by Hull in 1945. The procedure for RS SCA is similar with that for GCA, except that in RSSCA an inbred is used as a tester in the place of an open pollinated variety.

Main features of RSSCA

1 Used for the genetic improvement of polygenic character.

2 Selection is done on the basis of test cross performance.

3 An inbred is used as a tester.

4 Used for improving specific combining ability.

5 More effective with over dominance and less effective with incomplete dominance.

6 Used when a character is governed by non-additive (dominance and epistasis) gene action.

4. Reciprocal Recurrent Selection (RRS)

Reciprocal recurrent selection (RRS) is a cyclical breeding procedure designed to improve simultaneously the crosses of two populations from different heterotic groups by using both general and specific combining ability. In this procedure, genotypes from two populations are evaluated in reciprocal crosses and the best genotypes of each population are selected and recombined to give rise to improved population crosses. Inter population half-sib or full-sib progenies are used as evaluation units and intra population S_1 progenies as recombination units. This method was proposed by Comstock, Robinson and Harvey 1949.

Main features of RRS

1. Used for the improvement of polygenic characters.
2. Basis of selection on test cross performance.
3. Two heterozygous populations are used.
4. The two populations may be designated as A and B.
5. Used for improving a population both for gca and sca for specific character.
6. Equally effective with incomplete, complete and over dominance.
7. Used for the improvement of those characters which are governed by both additive and non-additive gene action.

Procedure

First Year: Two open pollinated populations A and B are used as base populations. Several plants are selected on phenotypic basis from both the population A and B. Each of the selected plants form the population A is crossed as male with several randomly selected plants from the population B used as female. Similarly population B used as male and population A as female. All selected plants are selfed, as well as test crossed. Test crossed seed and selfed seed from each selected plant is harvested separately.

Second Year: The test cross progenies of the selected plants of both the populations are evaluated in separate replicated trials. On the basis of progeny test, plants producing superior test cross progenies are identified.

Third year: Selfed seed from the plants producing superior test cross progenies of both the populations are planted in two separate crossing blocks. All possible inter crosses are made among the progeny of A plants and also among the progeny of B plants. Equal amount of seeds from all inter crossess A and B are mixed separately to carry to next generation of population A and B. This completes the original selection cycle.

Fourth Year: Population A and B are grown from the composited seeds from all intercross in block A and B separately.

Fifth Year: The operations of second year is repeated.

Sixth Year: The operation of the third year is repeated and this completes the first recurrent selection cycle.

The cycles may be repeated till desired result is obtained.

Systematic Representation of RRS

- Several phenotypicall superior plants are selected from the base the population A & B.
- Selected plants are self-pollinated.
- Each selected plants from A is test-crossed with several random plants from B and vice-versa.
- Test-cross progeny and selfed seed from each selected plant is harvested separately.
- Separate yield trials are conducted for test-cross progenies from population A and B.
- Superior progenies are identified.
- Selfed seed from plants producing superior testcross progenies planted separately for population A and B.

- All possible inter crosses are made separately among population A and B.
- Seeds from all inter crosses of a populations of A and B harvest separately
- Composite seed from inter crosses planted separately from A and B.
- Several plants selected in the population A and B.
- Selected plants self-pollinated.
- Each selected plants from A is test-crossed with several random plants from B and vice-versa.
- Test-cross progeny and selfed seed from each selected plant harvested separately.
- Separate yield trials are conducted for test-cross progenies from population A and B. Superior progenies identified.
- Selfed seed from plants producing superior testcross progenies planted separately for population A and B. All possible inter-crosses are made. Seeds from all inter crosses of a populations A and B mixed separately.
- May be repeated if required.

Advantages of Recurrent Selection

1. Recurrent selection is an efficient breeding method for increasing the frequency of desirable alleles in a population
2. Thus it is an important method of population improvement.
3. Provides greater opportunities for recombination to occur.
4. Helps in breaking repulsion phase linkages.
5. Maintaining high genetic variability in a population due to repeated inter mating of heterozygous populations.

Disadvantages of Recurrent Selection

1. This method is not used directly for the development of new varieties.
2. This is only used for population improvement.
3. Requires lot of selection, crossing and selfing work.
4. Permits selfing which leads to loss of genetic variability.

Table 10.1. List of Cross Pollinated Crops

Crop	S. Name	C. No.2n=
Maize	*Zea mays*	20
Pearlmillet	*Pennisetum glacum*	14
Sunflower	*Helianthus annus*	34
Safflower	*Carthamus tinctorius*	24
Castor	*Ricinus communis*	20
Coriander	*Corandrium sativum*	22
Fennel	*Foeniculum vulgare*	22
Rai	*Brassica nigra*	16
Niger	*Guizoia abyssinica*	30
Onion	*Allium cepa*	16
Cauliflower	*Brassica oleracea var. botrytis*	18
Cabbage	*B. oleracea var. Capitata*	18
Knol-khol	*B. oleracea var. gongylodes*	18
Brussels sprouts	*B. oleracea var. gemmifera*	18
Ashgourd	*Benincasa hispida*	24
Bittergourd	*Momordica charantia*	22
Bottlegourd	*Lagenaria siceraria*	22
Cucumber	*Cucumis sativus*	14
Muskmelon	*Cucumis melo*	24
Pumpkin	*Cucurbia species*	40
Ridgegourd	*Luffa acutangula*	26
Snakegourd	*Trichosanthes anguina*	22
Watermelon	*Citrullus lanatus*	22
Amaranths	*Amaranthus tricolour*	32
Beet leaf (palak)	*Beta vulgaris*	18
Spinach	*Spinacea oleracea*	12
Lettuce	*Lactuca sativa*	18
Beetroot	*Beta vulgaris*	18
Carrot	*Daucus carota*	18
Radish	*Raphanus sativus*	18
Turnip	*Brassica rapa*	20
Asparagus	*Asparagus officinalis*	20

11

Botany of Some Important Crops

1. Bread Wheat (*Triticum aestivum*)

Wheat has been traditionally grouped into the genera *Triticum* in the tribe *Tritiaceae* of the family Poaceae. Bread wheat, 2n=6x=42 is a hexaploid consisting of three distinct genomic complementary, namely, A, B and D, with the genome structure AABBDD. In the past there were 4 species, viz. Triticum aestivum, *T. durum, T. dicoccum* and *T. sphaerococcum*, under cultivation in India. *T. sphaerococcum* has now practically gone out of cultivation because of its low productivity and high susceptibility to diseases. The major wheat producing countries are China, India, USA, the Russian Federation, and Australia. These five countries together contribute more than half of the global wheat production. In India, Uttar Pradesh has registered the highest production followed by Punjab, Madhya Pradesh, Haryana, Rajasthan and Bihar. These top six states together contributed around 90 per cent of the total wheat production in the country

Floral Biology

The inflorescence of wheat is known as spikelets. Spikelets are alternately arranged on the rachis; each spikelet's has a small joint axis which bear two to nine flower known as florets. Close to the base of rachilla are two empty boat shaped bracts known as glumes. Above these inserted alternately on opposite side of the rachilla known as lemma. Opposite it's a thin membrane bracts known as palea enclosed within the lemma and palea are a single pistil and three steamens. The pistil consists of an ovary bearing two short styles each carrying a feathery stigma. At the base of the ovary are two membranous scales known as lodicules. Each stamen consists of a thread like filament carrying a pollen containing anther.

Emasculation and Crossing Techniques

Select the spike at the proper stage for emasculation. With the help of scissor one of the three of the basal and upper non-functional spikelet's are removed. With the help of scissor cut away one third of the spikeletes and lower spikelets are also removed. The top spikeletes is held with forceps and pulled downwards and upwards to remove the upper florets of the spikelets. Remove anther from the florets by inserting the forceps between the lemma and palea. Be sure that all the three anthers are removed without damaging stigma. The emasculated spike are immediately covered with paper bag and mention the date of emasculation. On the next day ear head selected from the pollen parent are used for crossing. Pollination must be done during morning time that stigma become receptive. Flower normally can be pollinated 2-4 days after pollination. The upper half of the glumes of the few medium spikelets is cut off and the ripened bright yellow anthers are rubbed on the styles of the emasculated florets and then covered

Breeding Objectives

1. Seed yield is identified as one of the main trait and many studies have been conducted on its inheritance.
2. The inheritance of seed yield may be expressed by morphological features of the plants, such as number of spikes, fertility of spikes and weight of grain.
3. The early maturing varieties are being preferred because it enables the wheat to escape some of the ill effects of the hot summer weather, drought and rust.
4. Breeding for resistance to lodging and shattering.
5. Breeding for resistance to biotic stresses, has become a major objective in wheat breeding.
6. Breeding of varieties resistance to abiotic stresses.
7. Breeding for development of grain quality.
8. Development of varieties for limited moisture contents.
9. Development of varieties for low gluten content

Important Varieties

Raj. 4120, Raj. 4079, VL 907, PBW 590, PBW 596, VL 907, MP 1202, Raj 6560, HD 2330, PDW 215, PVW 226, Raj 2184, HD 2189, Raj 3077, Raj 3765, Raj 4083, Raj 3777, DBW-187, DBW-222, Raj 4238, DBW 252, DDW 47, Karan Vandana (DBW 182), VL 832, VL 804, HS 365, WH 147, HW 741, HW 2044, Shresth (HD 268).

2. Rice (*Oryza sativa*)

Rice belongs to the genus *Oryza* of the tribe *Oryzae* of the family Gramineae with 2n=24 chromosome number. There are 24 recognized species, including the Asian and African cultivated rice. Cultivated rice can be divided in to *Oryza sativa* and *Oryza glaberrima*. *O. sativa* is grown world wide while *O. glaberrima* is grown only in parts of West Africa. *O. sativa* has been further divided in to Indica, *Japonica and Javanica*. *Indica* is adapted to tropical and sub-tropical Asia, While *Japonica* adapted to sub-tropical and temperate regions and *Javanica to* the equatorial region. Rice is the second most important food crop of the world. Rice occupies major parts of the plated area in South, Southeast and East Asia. The major rice growing states in India are, West Bengal, Bihar, MP, Chhattisgarh, Orissa, AP, UP, etc.

Floral Biology

The rice inflorescence is a panicle bearing single flowered spikelet's. The spikelet consists of rachilla, two sterile lemmas; a normal fertile lemma, palea and flower The lemma, palea and flower together known as floret. The sterile lemmas, lemma and palea are also called bracts. The tips of lemma and palea are known as apicules. The filiform extension of the keel of lemma is called awn. The flower consists of six stamens, one pistil containing one ovule and the perianth represented by lodicules. In rice anthesis commences shortly after emergence of panicle. Spikelet's at the tip bloom first and proceed downwards. Anthesis time in 8-10 am. Each spikelet remains open for 30 minutes and then closes. The anther dehiscence takes place immediately after the opening of the spikelet's. Receptivity remains for one day. Rice fruit is caryopsis.

Emasculation and Crossing Techniques

Emasculation is necessarily followed by controlled pollination. Emasculation is done during early morning between 6 to 8 a.m. in spikelet's. Emasculation should be over well ahead of the time of anthesis. Crossing techniques in rice differ based on the method of emasculation. Since maximum number of spikelets open on the 3rd or 4th day of anthesis, panicles of that stage are selected for emasculation.

Breeding Objectives

1. Collection, characterization and conservation of rice germplasm.
2. Development of high yielding rice varieties of different duration groups.
3. Improvement of agronomic features of the old varieties through induced mutagenesis.
4. Providing a high quality phenotype platform for drought tolerance and aerobic rice.
5. Studies on the diversity of pathogens and pests and, pyramiding genes responsible for major biotic stresses.

6. Adaptability and stability of yield.
7. Lodging and shattering resistance.
8. Improved grain quality like grain shape and size, texture of endosperm and quality of starch in endosperm, aroma and cooking quality (Basmati type), colour of kernel and milling out turn.
9. Breeding of varieties suited for direct seeding.
10. Breeding varieties for dry lands.
11. Breeding varieties for deep water conditions.

Important Varieties/Hybrids

Haryana Shankar Dhan-1, Bhutnath, Pusa Basmati-1, Pusa Basmati 6, Sahyadri –4, Vardhan, Vallabh Basmati-22, Narendra Usar Dhan, Vivek Dhan-62, Hybrid 6201, Anjali, Pusa Sugandh-5, Pusa 1121 (Pusa Sugandha-4), CR Dhan-501, Padma.

3. Pearl millet (*Pennisetum glaucum)*

Globally, pearlmillet popularly known as bajara. Pealmillet has diploid chromosome number 2n=14. The genus Pennisetum includes approximately 140 species. It is one of the largest genus in the tribe Paniceae. The section penicillaria includes 6 wild , 13 cultivated and 15 intermediate species. Clayton (1972) recognized six species,viz; *Pennisetum americanum, P.violaceum, P. fallax, P. stenostachyum and P. dalzielli.* In India, it is mostly grown in the arid and semi-arid regions, particularly in the northern part of the country, in the states of Rajasthan, Maharashtra, Gujarat, UP, and Haryana.

Floral Biology

The inflorescence is terminal, cylindrical and compact, spike having 10 to 50 cm length. Each spikelet consists of lower staminate and upper bisexual flower. The whorl consists of 30 to 40 bristles surrounded by 2 to 5 spikelet's. Each spikelet having two empty glumes and two inner glumes. The staminate flower is having single lemma and three stamens but does not have palea and lodicules. The bisexual and hermaphrodite flower has broad lemma, three stamens, palea, and single carple, style dividing into two branches. The ovary is unilocular containing single seed. It is highly cross-pollinated crop due to protogynous nature. The style begins to appear 2 to 3 days after emergence of spike. First at 1/3 rd portion and gradually downward, and style remain receptive for 1 to 2 days. Anther emerge after styles dry up. The anthers of bisexual flower appear 2 to 3 day before those of staminate flowers. The anther emergence starts from middle of the spike and proceeds upwards and downwards. Anthesis occurs throughout the day and night with the peak between 8.00 p.m. to 2.00 a.m.

Emasculation and Crossing Techniques

Hand emasculation is laborious and difficult due to small size of flowers and the late development of anthers in relation to stigma. Protogynous nature of the flower suggests the easy method of crossing without emasculation. In the lowermost part of the ear head the styles appear very later. This interval between appearance of styles and dehiscence of anthers is more particularly in lowermost region of the spike. Hence the spike emerged from the flag leaf is selected for emasculation and upper 2/3 rd or 4/5 portion of ear head is cut and remaining portion of the head is bagged immediately. It is always done in the morning after 2 to 3 days when styles appear, fresh pollen grains collected from protected male parent in petridish and dusted over the stigma of the bagged heads and rebagged immediately. The pollination may be repeated 2-3 times till the anthers of the same region show their appearance.

Breeding Objectives

1. Collection, characterization and evaluation of germplasm for yield and other important traits.
2. High yields is associated with more number of tillers, compact, long panicle, heavy grains and uniformi maturity.
3. Development of dual purpose hybrids, varieties, parental lines and novel genetic stocks with resistance to downy mildew, blast, ergot, smut and rust, etc.
4. Breeding for improved grain quality can be achieved by incorporating yellow endosperm to improve vitamin A content or white endosperm to improve protein content.
5. Breeding for drought tolerance can be obtained by evolving varieties having shorter duration, so that they can escape from drought.
6. Improve shelf life of pearl millet and reduce flour rancidity – understanding basic mechanisms
7. Strengthen research facilities and programmes on drought tolerance for development of parental lines and populations.
8. Breed varieties with more adventitious roots, high leaf water potential and high chlorophyll stability index.
9. Develop and popularize integrated crop production technologies and plant protection practices to major insect and diseases.

10. Breeding for alternate source of cytoplasm in male sterile lines.
11. Breeding for high forage value, high sugar content in the stem juice, increased leaf number with more breadth and digestibility.

Important Varieties

Varieties: PS B – 8, PSB 15, Mukta

Hybrids : HHB 45, HHB 50 GHB 30, GHB – 27, RHB-177 (MH-1486), RHB-177, RHB-154 (MH-1340), Pusa composite-334 (MP-334), Pro Agro 9555, PHB-2168, HHB 67 improved, HB 1, HB 3, BH 5, HHB 68, HHB 60, HHB 146.

Synthetic variety: ICMS 7703

4. Maize (*Zea mays*)

Maize which originated in Mexico is one of the most important cereal crop after rice and wheat globally. Belonging to the tribe Maydeae of grass family *Poaceae*, it is often referred as "Miracle crop" and "Queen of cereals". Maize is an allogamous crop and its wider morphological variability and geographical adaptability can be bestowed to its cross pollinated nature. With 2n = 20 number of chromosomes, maize is a monocious plant bearing staminate (tassel as male part) and pistillate (silk bearing ear or cob as female part) flowers separately. Being a protoandrous crop, tassels mature earlier than the female flowers and the anthesis starts from the central shoot of the tassel from the top and proceeds downwards. Maize is generally believed to have originated from teosinte more than 8,000 years ago in Central America, particularly South Mexico.

Floral Biology

Tassel is the male inflorescence comprising staminate flowers. It is much branched panicle in which spikelet's are arranged on both central axis of the panicle and on lateral branches. One sessile and other pedicelled spikelet's are in pairs. Staminate spikelets are two flowered, each with three anthers. The female inflorescence is known as cob or ear. The ear contains several vertical rows of densely crowded female flowers on the fleshy rachis. Generally, there are paired sessile spikelet's present on the rachis. Two flowers are found within each spikelet's. The lower flower is reduced to lemma and palea, both of which are membranous. The upper flower of spikelet's consisting of a pistil, 3 rudimentary stamens and two functionless lodicules. Pistil comprises three carpels, two extending to form the silk and the third forming the ovules. The ovary ends in a thread like structure, the silk, which is actually the modified style. Blooming starts in the tassel near the tip of the central axis and it proceeds downward. It takes nearly 14 days to complete flowering. The pollen grains remain viable for 24 hours. The pollen shedding begins 2 to 3 days before emergence of the silks.

Emasculation and Crossing Techniques

Manual emasculation is done that is known as detasseling. Detasseling is the removal of tassel from female parent. Detasseling is done when the tassel emerged out from the boot leaf, but before anthesis have shed pollen. Anthers take 2-4 days to dehisce after complete emergence. Only in few cases, the anthers start dehisces before its complete emergence. In such case detasseling should be done earlier. Detasseling is done every day from the emergence of tassel up to 14 days. Flower opening starts in the early morning and completes by afternoon. Pollen remains viable for a day. Emergence of the silk or style takes two to five days. Stigma remains receptive up to 14 days. On the, next morning, the pollen is collected in the tassel bag. Lift the bag from female inflorescence and dust the pollen on the cob. Bag the cob with tassel bag.Maize is cross pollinated crop.

Breeding Objectives

1. Develop genetically diverse, high-yielding maize inbred lines and hybrids.
2. Combining multiple-stress tolerance.
3. Enhanced nutritional quality.
4. Improve the efficiency of maize breeding operations.
5. Grain yield is the most important and complex economic trait.
6. The desired maturity varies depending on the cropping pattern for which the breeding is under taken.
7. Response to heat and drought.
8. Resistance to ear dropping.
9. Need to develop resistance varieties/ hybrids of the diseases like, seedling blight, root stalk, ear rot, smut, leaf blight, downey mildew, bacterial wilt and virus diseases.
10. The present emphasis is on the development of resistance inbred lines and hybrids.
11. Corn for milling.
12. Development of high yielding baby corn.

Important Varieties/Hybrids/Composites

Mahi Dhawal, Arawali Makka-1, Pratap Hybrid Maize-1 Pratap early Makka-3, Pratap Makka-4, Pratap Makka-5, Pratap Makka Chari-6, Pratap Hybrid Maize-2, Vivek-43, PMH-4, Pusa hybrid 5

Pop Corn Varieties: Jawahar Pop Corn, Pearl Pop Corn, Amber Pop Corn, VL Amber

Sweet Corn Varieties : Win Orange sweet corn, Priya Sweet Corn, Madhuri

5. Sorghum (*Sorghum bicolor*)

Sorghum is well known for its versatile uses. Millions of the people in the semi-arid tropical regions of the world eat sorghum as their staple food. Sorghum comes under the family of Gramineae and genus Sorghum. The basic chromosome number of sorghum is n= 10. Harlan and De Wet (1972) developed a simplified, informal classification useful to plant breeders. Bicolor (B), Guinea (G), Caudatum (C), Kaffir (K) and Durra (D). It is believed that the centre of origin of sorghum is north-eastern Africa. It is mainly cultivated in the USA, India, Africa, Latin America and Australia. In India major sorghum growing states are Maharashtra, MP, AP, Rajasthan and Karnataka.

Floral Biology

The inflorescence of sorghum is the panicle. The panicle may be short and compact or loose and open. The central axis of the panicle, is rachis. The raceme always consists of one or several spikelet's. One spikelet is always sessile and the other pedicellate. The racemes vary in length according to the number of nodes and the length of the internodes. Glumes also vary from quite hairy to almost hairless and are hard and tough. There are two lemmas, each one with delicate white tissue easily overlooked at a casual glance. The lower lemma is elliptic or oblong, about equal in length to the glumes. The upper lemma is shorter, more ovate and may be awned. There are two lodicules and a palea. Sorghum has two pistils and three stamens. Each fluffy stigma is attached to a short stout style extending to the ovary. The anthers are attached to long thread like filaments.

Emasculation and Crossing Techniques

Flowering in sorghum begins at the tip of the head and flowers successively downwards over a period of 4 to 5 days. The panicles that open the next day are selected for emasculation. The top portion of the panicle and very young spikelets at the base are removed. After this, the staminate florets are also trimmed off. Now the glumes of each spikelet are opened with a medium sharp forceps or needle. The spikelet is held between the thumb and the fore finger. Anthers are then pushed out by applying pressure through the palea. This method of emasculation is called the hand with a rotating movement. Another method is known as the hot water method of emasculation. In this method, the panicle with the spikelet's that will open the next day are immersed in water at temperature of about 47°C to 49°C for three to five minutes. This temperature kills the pollen but not pistil. The emasculated panicles are enclosed in paper bags. The next morning, pollen is collected from the selected male parent in the paper bags, which was previously bagged and dusted on the emasculated panicle and then again covered by the paper bag.

Breeding Objectives

1. High grain yield.
2. Wide environmental adaptiveness.
3. High forage yield.
4. Resistance / tolerance to insects like shoot fly, stem borer, gall midge etc.
5. Resistance to biotic stresses like grain mold, downy mildew, rusts, leaf blight, leaf spots etc.
6. Breeding for non-lodging.
7. Appropriate time to maturity.
8. Good threshability.
9. General attractiveness.
10. Large head size, good head exertion.
11. Good tillering, with heads on all culms maturing the same time.
12. Dual purpose genotypes with high grain and fodder biomass potential per unit time.
13. Good seed size and number.
14. Breeding for special traits like sweet sorghums and striga resistance.

Important Varieties/Hybrids

Hybrids: CHS 1, 5, 6, 9, 13, 14.

Varieties: CSV 11, 13, 15, Swati, M 35-1, CSV 14 R

6. Mustard (*Brassica campestris*)

The oil seed brassica commonly called as rapeseed and mustard. The genus *Brassica* includes about 150 species. The diploid Brassica species have n=8, 9 and 10 chromosomes. *B. campestris* has the widest distribution and its primary centre of origin is Himalayan region and secondary centre of origin is European Mediterranean area. Brown sarson originated in eastern Afghanistan and western part of India.The yellow sarson originated in eastern India and *B. juncea* originated in China. *Eruca sativa* is a native of Southern Europe and North Africa. Mediterranean region is the place of origin of *B. napus* whereas, *B. carinata* originated in north east Africa. Indian mustard is cultivated in the states ofAssam, Bihar, Gujarat, Rajasthan, MP, UP, Haryana, J.& K. and Orissa.

Floral Biology

The inflorescence of mustard is racemose and flowering is indeterminate type. Flower is regular, bisexual and hypogynous with four free sepal into two whorls. Flowering period may last 2-3 weeks. Normally stigma is receptive for three days after opening of flower. All the pollen are shed on the day flowers open, provided the dry weather occurs. The flowers form a funnel shaped structure during the evening. The pollen can be stored for 4-5 weeks without the loss of viability. The syncarpous ovary develops into pod with two carpel separated by a false septum called repcom.

Emasculation and Crossing Techniques

The bud is opened with the help of fine and pointed forceps. Six anthers are removed. Generally emasculation is done during day time but evening is preferred. Protect the emasculated bud from the foreign pollen by using plastic bags for up to one week following pollination. Collect the pollen from the male parent by using hair brush. Dust the pollen on the tip of the stigma of emasculated bud gently. Collected pollen can be stored upto five weeks without the loss of viability. Covering the emasculated bud after pollination. Care must be taken to clean the hair brush between the different pollination.

Breeding Objectives

1. The ultimate objectives of the Brassica breeding programme are to develop varieties that produces higher seed yield.
2. An improvement in oil content will ultimately increased the oil yield. It is easier to increase oil content than seed yield, since oil content is controlled by small number of genes.
3. Incorporation of yellow seed colour in commercial varieties will further increase oil and protein content.
4. Need to reduce anti-nutritional factors by breeding.
5. The plant with compact branching is preferred.
6. Early maturing varieties are needed for use in various multiple cropping sequences.
7. *B. campestris* varieties maturing in less than 100 days are required to escape from frost damage.
8. Short-duration rapeseed varieties are suitable to avoid late-season drought stress.
9. Brassica yields suffer because of many abiotic stresses associated with low temperatures, frost, drought, alkalinity and salinity.
10. A large number of fungi and viruses are pathogenic to oilseed brassicas.

Important Varieties

PR-45 (Pusa Bold), PCR-7 (Rajat), Sej-2 (Pusa Agrani), RH-30, RLM 619, RH-819, RK-9902 (Maya), RB 9901 (Geeta), RH-9304 (Vasundhara), RGN-13, T-59 (Varuna), Bio-902 (Pusa Jai Kishan), RN-393 (Arawali), Giriraj, DRMR IJ 31 RH-749.

7. Groundnut (*Arachis hypogeae*)

Groundnut having chromosome number 2n=40 and belongs to the family of Leguminoseae. Brazil is the origin place of groundnut.Groundnut is grown largely in India, China, USA, Senegal, Indonesia, Nigeria. Myanmar, Brazil and Argentina.In India, it is cultivated in AP. Gujarat, Karanataka, Tamil Nadu, Maharashtra, Rajasthan, UP, MP, and Orissa. The genus Arachis is subdivided in to the following seven sections. Arachis, Erectoides, Rhizomatasae, Extranervosae, Triseminate, Ambinervosae and Caulorhizae.

Floral Biology

The inflorescence is either a single or compound in the axils of cataphylls, Flowers are complete, zygomorphic with reduced pedicel and bracteates. Calyx has five lobes in two groups of one and four. The gamosepalous calyx forms a green coloured tube. Petals are five in number and bright yellow in colour. The keel is pale yellow and closely wraps around the stamens and the upper part of the style and stigma. The stamens are ten in number, out of which two are sterile and remaining eight are dimorphic, monoadelphous condition, forming a stamina tube. The anthers are oblong and globose alternatively. The ovary is situated at the base of the hypanthium. It is superior and normally has two-four ovules and rarely six, it is about 1.5 mm long. The style is long and fillform. It has hairs on its distal portion. The stigma is club shaped and usually level with or slightly above the anther. The stigma become receptive about 24 hrs. before anthesis and its receptivity prolongs for about 12 hrs after anthesis. Flowers are open between 6-8 am but may be delayed by low temperature. Generally anthers may dehisce 7-8 hrs .before flower opening.

Emasculation and Crossing Techniques

Emasculate the young buds in the afternoon or is carried out between 1.30 to 4.30 pm or evening depending upon the prevailing environmental condition. Once the bud is selected, all other buds at that nodes are removed with the fine forceps. Expose the bud by pulling the leaf gently. The bud is held gently between the thumb and first finger of the left hand using forceps held in the right hand, The single sepal opposite the standard petal is pulled down. Open the standard carefully with forceps and is held back by thumb and index finger. Expose the anthers by pulling the bud outwards with the help of forceps. Remove all eight fertile and two sterile stamens. The standard wing and keel petals usually return to their normal position after emasculation to cover the style and stigma. For pollination a healthy flower from a pre-identified male parent is removed by breaking the hypanthium. Ditch the calyx, standard and petals. Tease the anthers with needle and dust pollen on the stigma of the female flower with the help of camel hair brush. Labeling is done.

Breeding Objectives

1. To create variability by various breeding approaches.
2. To develop high yielding varieties.
3. To develop early maturing varieties.
4. To evolve abiotic stress tolerant varieties.
5. Breeding high yielding bunch ground nut with dormancy suitable for dry land conditions.
6. Breeding varieties for high shelling percentage > 75% and high oil content > 50%.
7. Breeding disease resistance varieties for early leaf spot, late leaf blight, tikka leaf spot, rust, collar rot, and dry root rot.
8. Breeding for pest resistant varieties like,red hairy caterpillar, bihar hairy caterpiller, leaf minor, white grub, jassids, aphids and thrips.
9. Breeding short duration (85 days) varieties suitable for irrigated conditions

Important Varieties

Tirupati-2, ICGS-37, TKG-19A, BRG-12, HMT-10, TG-17, TG-26, ICGS-1, ICGS-11, ICGS-44, Samrath, JCC-88, G-8808, RG-382 (Durga), HNG-10, DRG-17, RG-141, Mallika.

8. Sesame (*Sesamum indicum*)

It is an ancient oil seed crop of tropics and warm sub-tropics. India, and Ethiopia (Africa) are considered as centre of origin. s Sesame belongs to the family of pedaliaceae and having chromosomes number 2n=32. The genus *Sesamum*, contains over 30 species, of which only *S. indicum* is cultivated. The wild species *S. angusstifolium* and *S. radiatum* are also cultivated in Africa. Sesame is cultivated in India, Pakistan, Africa, China, Mexico, Iran, Iraq etc. Statewise, Rajasthan, Orissa, Maharashtra, MP, Gujarat, UP, AP, Karnataka, Tamil Nadu and West Bengal are the major producers of sesame.

Floral Biology

Inflorescence is cymose type. Flower is pedicellate, hermaphrodite, zygomorphic, complete, hypogynous, ebracteate. Five sepals, gamosepalous, sepals' lanceote, acute. Five petals, gamopetalous, 2- lipped, corolla 5-lobed, the upper lip consists of 2 equal sized lobes while the lower lip includes 3 petals, out of which central one is the largest. Androecium 4 stamens, didynamous, epipetalous, instead of 5th usually a staminode is present. Anthers 2-celled (dithecous), stamen sagitate, introrse and dehisce by longitudinal slit. Gynoecium two carpals (bicarpellary) or four carpals, syncarpous, ovary superior, placentation axile, style filiform, stigmas 2 and lobed.

Emasculation and Crossing Techniques

For crossing, emasculation is easily performed by pulling off the corolla from the buds on the day before opening. The gamopetalous corolla is smoothly slipped out by a gentle pull. As the stamens are epipetalous, they get themselves simultaneously removed along with corolla. Flowers open at dawn. The pollen grains are shed shortly thereafter, remain viable for about 24 hours. The stigma is usually receptive for 24 hours after flower opening. Although, the best times for emasculation and pollination are the afternoons and mornings respectively, these processes can be performed at the same time. For pollination, healthy and robust flowers of male parent are collected by smooth slipping of the corolla column. The anther with fresh pollens are exposed by splitting open the tip of the corolla, and brushed on stigma of the emasculated flowers.

Breeding Objectives

1. Higher yield is the first and most important objective.
2. Optimum number of capsules per unit area.
3. Maturity, early type, medium and late, as per local requirements.
4. Improved plant architecture.
5. Potential to respond with higher seed yields to applications of fertilizers.
6. Need to develop non-shattering types.
7. Seed should be large, well filled, shape and colour.
8. Resistance to diseases and pests and indehiscent capsules are the major objectives in this crop .
9. Resistance to water lodging, drought, salinity and other abiotic stresses.
10. The development of varieties with low or zero anti- nutritional factors like oxalic and phytic acids needs attention for its value addition.
11. Increase in oil content is also one of the important components in varietal improvement of this crop.

Important Varieties

TKG-306, 308, SWB-32-10-1 (Savitri), Jawahar Til-12 (PKDS-12), RT 46, 103, 125, 127, 346, 351, Smarak, Shubra, DSS 9, akt 64, 101, Sekhar.

9. Soybean (*Glycine max*)

Soybean, golden nugget of the orient, recognized worldwide as a miracle crop. Taxonmoically, the genus glycine is a member of the family *Leguminoseae,* sub *family Papilonoideae,*tribe phaseoleae and sub-tribe *Glycininae.* The subspecies *Glycine* is composed of 16 wild perennial species. Twenty chromosomes of the haploid complements (2n=40) consists of 14 intermediate and four small chromosomes. Progenitors of soybean are *G. ussuriensis*, *G. tomentella*, *G. tabacina*, *G gracilis*. *Glycine* max is originated from *G. usuriensis* and *G. tomentoll*a. Centre of origin of soybean is china. It is growing in the countries like, USA, Brazil, China, Argentina and India. The soybean growing states in India are MP, Maharashtra and Rajasthan.

Floral Biology

The raceme type of inflorescence is found in soybean. The flowers are borne on axillary racemes and terminal racemes. The individual flowers are arranged on a stem-like branch called a peduncle. The flowers' may occur in clusters of 3-15 per node. Axillary buds that develop into flowers expand first towards the base of the main stem and then, after progress, towards the tip. The initiation of flowers takes place generally at the fourth node. Many flowers shed without forming pods. Soybean flowers are 6-7 mm long, bisexual, pedicellate, zygomorphic, pentamerous, bracteate and typically papilionaceous with a tubular calyx of five unequal lobes. Calyx is persistent and attached with pods. The corolla comprises a posterior banner or standard petal, two lateral wing petals and two anterior keel petals. The keel petals are attached to each other but not fused. The petal colour is either violet or white. The 10 stamens (androecium) are in diadelphous pattern. Nine stamens are fused and develop in a tube around the pistil. The tenth stamen remains free. Anthers are bilobed, dorsifixed and introse.The single pistil (gynoecium) is unicarpellary, superior, unilocular, curved style towards the free posterior stamens. The pistil has one to four campylotropous (curved) ovules with marginal placentation.

Emasculation and Crossing Techniques

Soybean contains small floral parts due to which artificial cross-pollination becomes a tedious job. Hence, special care should be taken for a successful hybridization programme. Parental cultivars may be grown in pots or in fields (pair rows) with sufficient spacing at different intervals. This way, crossing between early and late types of cultures becomes possible and a sufficient number of crossed pods or seeds can be achieved. Flowers are prepared for crossing just as the floral bud is swollen and the corolla is visible through the calyx. The other buds should be removed gently with forceps without any injury to the desirable bud. In the process of emasculation, the very first operation is to remove the sepals with the help of forceps by pulling down the lobes one by one.In this way, the corolla is exposed, then, with the help of forceps, catching them and pulling upward, all the petals are removed. Simultaneously, 10 anthers are also removed usually. However, if some stamens are left, with the help of forceps, gently keep them out.

Breeding objectives

1. High yield is one of the most important objectives in soybean as in any crop.
2. Varietal development programme involves suitability of maturing period for a given environmental.
3. Breeding for resistance to lodging. The dwarf, determinate and strong stem characters are required for lodging resistance.
4. Resistance to shattering. It required retaining seeds in pod after maturity.
5. Breeding for forage type of soybean.
6. Breeding for vegetable type.
7. Resistance to abiotic stress like, water stresses, temperature, salinity etc.
8. Resistance to diseases viz; seedling and stem rot, collar rot, rust, soybean yellow mosaic virus, pod blight, bacterial pustules, soybean mosaic virus, myrothecium leaf spot etc.
9. Resistance to inspect-pest, like, field cricket, seed maggot, cut worms, blue beetle, stem fly, girdle beetle, jassids, thrips, white fly, leaf minor etc.

Important Varieties

NRC 2, NRC 12, NRC 7, NRC 37 (Ahilya 4), Pb 1, PK 1042, PK 1029, PK 472, Pusa 20 (DS-74-20-2), VL-Soya 21, MAUS 47, MAUS 61, MAUS 61-2, MACS 450, JS 93-05, JS 94-41, Pusa 16, 22,23.

10. Chickpea (*Cicer arietinum)*

Chickpea belongs to the family of Leguminosae and genus *Cicer*. The genus *Cice*r comprises of about 40 species. Chromosome number of all the species is 2n=16. Chickpea is cultivated in almost all parts of the world covering more than 50 countries spread over Asia, Africa, Europe, Australia, North America and South America continents. In India, it is cultivated in MP, Rajasthan, Maharashtra, UP, AP, Karnataka, Chhattisgarh, Bihar and Jharkhand. Vavilov

(1926) mentioned South-west Asia and the Mediterranean as primary centre of origin and Ethiopia as a secondary source of origin.

Floral Biology

Flowers are solitary axillary. They are pedicellate, hermaphrodite, complete, zygomorphic, perigynous, polypetalous with a axillary aestivation. Calyx tube is oblique, gamosepalous and densely covered with hairs. Corolla is veined pinkish, purplish or red or white. Stamens are ten in number and diadelphous (9+1) in condition. The anthers are two-celled uniform bursting longitudinally. Ovary is superior with one or two ovules having slightly bent and blunt stigma. The flowers open between 9 to 10 am and proceed up to 3 pm. The fruit is a turgid pod containing one or two seeds. Anther dehiscence takes place in the bud a day before the flower opens.

Emasculation and Crossing Techniques

The emasculation is very difficult in chickpea because of small size of the flower. The flower bud which is going to open the next day should be selected. The bud is gently held between thumb and fore finger of the left hand and the standard is just turned above with the help of forceps. Take out all the ten anthers by exposing wings and keel to ensure that all the ten are removed and no anthers bursts after that the emasculated bud is covered with the paper bags to avoid cross pollination. Flowers from the male parents should be picked in the morning of anthesis and dust yellow colour pollen grains of half -open flowers on the receptive stigma of the emasculated bud. Pollination should be done in the morning between 8am 12 noon. After pollination the flower bud is again covered.

Breeding Objectives

1. Yield potentiality can be increase by breeding cultivars responsive to fertilizer and irrigation.
2. Breeding of cultivars suited for double cropping.
3. Breeding for resistance to biotic stresses will help to stabilize chickpea production.There are many types of biotic stresses which deprives the yield production. Ascochyta blight, wilt and root rot (diseases), pod borer and leaf minor (pests), cyst, root knot, root, lesion nematodes.
4. Breeding for resistance or tolerance to abiotic stress is an important objective because these result in high yield instability. The major abiotic stresses are drought, salinity, cold and frost.
5. Produce bold seeded varieties.
6. Improvement of quality traits
7. Development of double podded varieties.

Important Varieties

RSG-973, Aruna, Gauri (GNG-1499), Aparna, GNG-1488, Gangaur (GNG-1581), RSG-974, RSG-44, GNG-1292, GNG-663 (Vardan), CSJD-884 (Aakash), RSG-895 (Arpita), RSG-945 (Aasha), RSG-888 (Anubhav) GNG 2144, GNG 1958, GNG 2171, GNG 2207, GNG 2299.

11. Pigeon Pea (*Cajanus cajan*)

Pigeonpea bclongs to the genus Cajanus, sub-tribe cajaninae, tribe Phaseoleae and family Fabaceae. Several species are listed under this genus, namely, Bicolour, *Luteus, Indicus, Cajan and Cytisus.* Pigeonpea is originated in India and spread to South–East Asia. Pigeonpea is an important pulse crop, grown in more than 22 countries including India, Myanmar, Malawi, Tanzania, Uganda, and Kenya. About 85% of the pigeon pea is grwn in six states of India namely, Maharashtra, Karnataka, MP, AP, Gujarat and Jharkhand and other states includes, UP, Orissa, Tamil Nadu, Chattisgarh, Rajasthan, Punjab, Haryana, West Bengal and the Northern-Eastern States.

Floral Biology

Inflorescence is small, terminal or axillary or both. Flowers with long peduncle are borne on terminal. The raceme inflorescence forms a terminal. Flowering proceeds acropetally (in the direction of the apex) in the raceme. A cluster of flowers is formed at the top. Papileonaceous, bisexual and zygomorphic flower. Calyx is five-lobed, gamosepalous. The upper lobes are paired, free or partly free, whereas the lower one is the longest. The corolla is zygomorphic and bright yellow. The imbricate petals are of three prominent types, that is, the standard, wings and keel. The keel covers the stamen and pistil. Ten stamens are arranged in diadelphous condition, 9 + 1. The ovary is superior, sub-sessile, flattened and dorsiventral with a long style. The style is filiform, unturned beyond the middle region and glabrous. Ovary is monocarpellary and unilocular with marginal placentation of two to nine ovules.

Emasculation and Crossing Techniques

The mature buds (1-2 cm size) that open one or two days later are selected .It is desirable to select two buds on a twig. 2 to 10 buds can be emasculated on an inflorescence.The rest of the flower buds and pods in the branch are removed. The bud is held firmly between thumb and forefinger of the left hand without damaging it while removing the sepals covering the keel. Then open the standard petal in an oblique position along the top of the bud by inserting the tip of the forceps at the base of the keel. Move the forceps upward to the tip of the standard and press the bud with a slight pressure until it opens. This makes the anthers visible. Remove all anthers from the staminal column. Therefore, pollination should be carried out. Morning emasculation followed by immediate pollination showed highest pod.

Breeding Objectives

1. Evolution of long duration high yielding varieties suitable for rain fed to replace the local land races.
2. To breed short duration (105 days) varieties suitable for irrigated / mixed crop with ground nut.
3. Breeding for bold grain type with desirable seed coat color.
4. Breeding for vegetable type. Green pods with bold seeds are used as substitute for green peas in some areas are perennial types.
5. Breeding of high yielding varieties resistance to sterility mosaic, phytophthora blight, wilt, alternaria blight, pod fly and pod borer in pigeonpea are required to stabilize the performance of seed yield.
6. Breeding of high yielding varieties tolerant to high and low temperature, water lodging, drought, acidic and salinity in order to bring more area under pigeonpea cultivation.
7. Breeding for high protein content and quality varieties.
8. Developed varieties suitable for inter-cropping, relay cropping, and multiple cropping and intensive cropping are important objectives in this crop.

Important Varieties

Varieties : UPAS 120, VLA 1, PAU 881, PA 6, CO 6, Pant A 291, Pusa 291, Pusa 992, GT 104, ICPL 87119, Pusa 9, MVB 106, IPA 203, PAU 881, MPV 106, IPH 15-3, CoRG 2012-25, WRGE 93.

Hybrids: ICPH 8, PPH 4, CoPH 1, 2, AKPH 4101, 2022, GTH 1.

12. Mungbean *(Vigna radiata)*

Mung bean is an important legume crop cultivated in the cereal based farming systems in Asia. Mungbean is classified in the family of Leguminosae, subfamily papilionaceae, genus *vigna* and sub-genus *certatotropics.* The genus vigna has about 150 species, of which 22 are native to India and 16 to Southeast Asia. Mungbean has a diploid chromosome number 2n=2x=22. Mungbean probably originated in India or the Indo-Myanmar region of Southeast Asia.

Floral Biology

Inflorescence bears small flowers in capitate clusters on the end of long having peduncles. The flowers are hermaphrodite, zygomorphic, calyx comprises five sepals, three large and free two small and fused, five petals (one standard two wings and two keels united) and ten stamen (9+1) diadelphous. The gynoecium is monocarpellary with superior unilocular sessile ovaries. The stigma is oblique and hairy. Flower open between 6 to 8 am and remain open till 11 am. Anthers start dehiscing by 9 pm and completely dehisced by 3 pm.

Emasculation and Crossing Techniques

Select the appropriate bud for emasculation since the fertilization in the mung bean takes place in the bud stage therefore the age of the appropriate bud is important. The appropriate bud is identified from the appearance of the corolla just above the calyx. Remove the immature anther with the help of forceps. Remove all the buds except those which are to be emasculated. Collect the anthers from the newly opened flowers of the male parent and gently put on the surface of the stigma of emasculated flower. Tie a tag to it.

Breeding Objectives

1. Concentration of combinations of genes for genetic potential to produce high seed yield.
2. Development of varieties suitable for different growing seasons.
3. Development of varieties suitable for high temperature.
4. Development of varieties resistant to damping off, seedling blight, root and stem rot, leaf spot and powdery mildew.
5. Breeding for insect resistant varieties, particularly for pod borer, white fly, aphid, leaf crinkle and leaf eating caterpillar.
6. Breeding to increase protein content and amino acid balance.
7. There is needed to develop the varieties for non-shattering types.

Important Varieties

MUM 2, ML 337, Pusa 105, Pant Mung 1, 2, 3, Gujarat 1, RMG 62, K 851, Narendra Mung 1, PDM 11 (Basanti), PDM 54 (Moti), PDM 139 (Samrat), IPM 99-125 (Meha), IPM 02-14, IPM 02-3, MH 421, IPM 410-3, Pant-5HGM-16, P 672, GM 07.

13. Cotton (*Gossypium* spp.)

Cotton belongs to the genus *Gossypium* of the family Malvaceae and tribe *Gossypieae*. The chromosome number of Asiatic cotton is 2n=26 while that in the New World cotton is 2n=52. Cotton composed of about 50 diploid and tetraploid species, but word fiber is produced from four species, namely, *Gossypium arborecem, Gossypium herbaceum, Gossypium barbadense* and *Gossypium hirsutum. G. hirsutum* is native to Mexico and parts of Central America and grown in upland sites, called upland cotton. *G. barbademe* is considered to have originated in South America and was adapted to sea island. Cotton is grown in over 80 countries, In India, Punjab, Haryana, Rajasthan, Gujarat, Maharashtra, MP, AP, Karanataka and Tamil Nadu.

Floral Biology

Cotton produces two types of branches. The lower branches are monopodial and upper branches are sympodial. The monopodial is vegetative and sympodial is fruiting. Flowers are extra-axillary, terminal or solitary, creamy, red or purple in colour. Three triangular bracts forming squares surround the cotton. The flower contains an open corolla with five petals, a staminal column bearing clusters of stamens and forming a tube that encloses the style. The pistil consists of 3-5 carpels. The ovary develops into five-loculed capsules or balls. Each locule has 7-9 seeds.

Emasculation and Crossing Techniques

Flower buds, which are likely to open on the following day, are selected for emasculation. The corolla is cut away with the help of scissors and the staminate column is removal of anthers between 3-6 pm and covered with a tissue paper bag to prevent the chance of cross pollination. Pollen grains of desirable male parent are collected and crossing is done by dusting the pollen grains on the emasculated stigma. Fertilization is completed about 30 to 40 h after pollination. About 18-25 seeds may be obtained from a cross ball.

Breeding Objectives

1. Breeding for high seed yield (more bolls and bigger bolls).
2. Breeding for high lint production.
3. To improve fiber yield of a cotton plant, which is determined by number of boll, size of the bolls and percentage of lint.
4. To develop cultivars that set a high percentage of five-lock bolls, are superior in yielding ability.
5. The spinning performance and quality of cotton fiber is associated with length, strength and fineness of the fibers,need to develop such type of varieties/hybrids.
6. Breed rapid fruiting and early maturity cultivars that reduce losses duc to disease and insect pests, facilitates harvesting with a mechanical picker,

increase production efficiency by reducing inputs of fertilizer and protective chemicals and irrigation water.

7. To develop early maturing varieties/ hybrids.Earliness is influenced by time of flowering initiation, rate of development of new flowers and the time period required for the bolls to mature.
8. To develop varieties for abiotic stresss like, drought, water stagnation, salinity and high temperature.
9. To develop cotton varieties/ hybrids without gossypol in the boll.
10. Breed varieties/ hybrids resistant to diseases like, bacterial blight, fusarium wilt, verticillum wilt, root rot, grey, mildew etc. and insect-pest like, whitefly, bollworm, stem weevil and red cotton bug.

Important Varieties

Jayadhar, Hybrid-4, RST-9, Varalaxmi, Suguna, H-6, DCH-32, LRA-5166, Dh-7, NHH-44, LRK-516, AKA 8401 and MCU-5, RG-18, RS 2013), Hybrid 4, Bikaneri Nerma, RAJ HH-16 (Maru Vikas).

12

Heterosis, Inbreeding and System of Mating

The commercial exploitation of heterosis started only after the rediscovery of Mendel's law of inheritance and consequent understanding of the theoretical basis of hybrid vigour. Heterosis is considered as a major revolution in classical plant breeding era. Utilization of heterosis has substantial influence in genetics to provide hybrid vigour, consequently high yielding hybrids in different crops and plants. Yield increase from heterotic hybrids due to the expression of heterosis is up to 30% superior to conventional varieties. The term heterosis was first used by Shull (1914) and represents the increase or decrease in fitness and vigour of the F_1 hybrid over the mean or mid parent values, produced by crossing two genetically different individuals for one or more traits. Generally, positive heterosis is desirable but in some cases like maturity duration and presence of toxic substances, negative heterosis is condiderable as desirable. Heterosis can be estimated in three different ways. When the heterosis is estimated over mid parent, it is called as relative heterosis or mid parent heterosis. Estimation of heterosis over better parent out of the two parents involved in the cross is referred as "Heterobeltiosis", the term which was coined by Fonseca and Patterson (1968). "Standard Heterosis", often called as useful heterosis which was coined by Meredith and Bridge (1972) and denotes the superiority of F_1 over the standard commercial check.

Important features of Heterosis

1. Heterosis leads to superiority in adaptation, yield, quality, maturity, disease resistsance and general vigour over its parents.
2. Generally positive heterisis is considered desirable, but in few cases negative heterosis is also desirable for the characters like flowering, maturity, plant height etc.
3. Heterosis is confined only to the F_1 generation.
4. Heterosis declines in F_2 and subsequent generations, as a consequence of segregation and recombination.
5. Heterosis once identified, can be easily reproduced in a definite environment.
6. It has positive association with SCA.

7. The magnitude of heterosis is associated with heterozygosity.
8. Deleterious recessive genes are marked by favourable effect of dominance genes.

Genetic Basis of Heterosis

Based on gene action and interaction, a number of hypothesis have been proposed to understand the heterosis.

1. Dominance hypothesis

The concept of dominance hypothesis was first proposed by Davenport (1908) and later supported from the researchers of Bruce (1910), Keeble and Pellow (1910) and has been discussed by Whaley (1944), Gowen (1952) and Stringfield (1964). According to this hypothesis, heterosis is the result of the superiority of dominant alleles. When recessive alleles are deleterious; the deleterious recessive genes of one parent are hidden by the dominant genes of another parent and the hybrid exhibit heterosis. Both the parents differ for dominant genes. For example, in a cross of two inbred lines A and B having the genetic constitution of one parent is AAbbCCdd and that of another as aaBBccDD. A hybrid between these two parents will have four dominant genes and exhibit superiority over both the parents which have two dominant genes each. Thus heterosis is directly proportional to the number of dominant genes contributed by each parent.

AABBccdd X aaBBccDD
(Parent 1) (Parent 2)
AaBbCcDd (Hybrid)

2. Over dominance hypothesis

The theory of over-dominance for the genetic explanation of heterosis have been advocated by Shull and East in 1908 and supported by Stadler (1939), Gustafson (1938) and Hull (1945). This hypothesis is the resultant of superiority of heterozygote over its both homozygous parents. Thus heterosis is directly proportional to the heterozygosity. The superiority of heterozygote over both homozygotes may arise either due to production of superior hybrid substances in heterozygote that is completely different from either of the homozygous products or due to greater buffering capacity in the heterozygote resulting from cumulative action of divergent alleles.

Estimation of Heterosis

1. **Average heterosis** : When the heterosis is estimated over the mid parent.

   Average heterosis = $[(F_1\text{-MP})/MP]\times 100$

2. **Heterobeltiosis**: When the heterosis is estimated over the superior or better parent.

   Heterobeltiosis = $[(F_1\text{-BP})/BP]\times 100$

3. **Standard heterosis**: When heterosis is estimated over the standard commercial check variety.

   Standard heterosis = $[(F_1\text{-CV})/CV]\times 100$

Inbreeding

Inbreeding refers as mating together of individuals that are more closely related by ancestry than would be expected under random mating.

Inbreeding Depression

Inbreeding depression may be defined as the reduction or loss in vigour and fertility as a result of inbreeding. The degree of inbreeding of an individual is expressed as inbreeding coefficient.

$$\text{Inbreeding depression} = \frac{F_1\text{-}F_2}{F_1}\times 100$$

Features of Inbreds

1. Inbred lines are produced by continuous selfing (about 5-6 generation of selfing).
2. Inbred lines are homozygous.
3. An inbred line consists of individuals with the same genotype.
4. They are maintained and developed by repeated selfing of selected plants.
5. Inbred lines are developed from a variable source population like, open-pollinated variety, synthetic, a single cross or a double cross.

Effects of Inbreeding

1. Due to inbreeding, there is a general reduction in vigour of the population.
2. The reproductive ability of the population decreases rapidly.
3. The population rapidly separates into phenotypically distinct lines.
4. Appearance of lethal and sub lethal alleles.
5. Increase in homozygosity.

Procedure for production of inbred lines

1st **year** : i) Plants with desirable phenotypes are selected. ii) Plants should be vigorous and disease free iii) Selected plants are self-pollinated by bagging.

2nd **year** : i) 30-40 or more selfed plants are space planted. ii) Best plants are selected from best progeny and self-pollination.

3rd – 6th yr: i) Same as 2nd year but number of generations of self pollination increases and selection is primarily done among the progenies.

7th **year**: i) Individual plant progenies that are nearly homo-zygous and homogeneous are harvested. ii) selfing may be discontinued and maintained by sib-pollination

Methods used for improvement of inbreds

- Pedigree selection
- Backcross method
- Convergent improvement
- Gamete selection
- Somatic hybridization
- Somaclonal variation
- Genetic engineering

Applications of Inbreds in Crop Improvement

1. Inbred lines are the genotypes that are used as parents in the production of hybrid and synthetic varieties in the breeding of cross-pollinated species.
2. The success of a crop breeding program relies on choice of the best parents possessing complementary and desired traits.

Systems of Mating

Various schemes used for mating of individuals are known as systems of mating.

1. Random mating

Each female gamete have equal chance to mate with any male gamete and the rate of reproduction of each genotype is equal i.e. without selection. In such a situation, gene frequencies, variance, correlation between relatives or prepotency remain constant. In random mating with selection changes the mean of the characters, increases the frequency of alleles for which selection is made. Increase the variability when selection is done for rare allele and reduce when predominant allele is selected. Useful for progeny testing, production and maintenance of synthetic and composite varieties and production of poly cross progenies.

2. Genetic assortative mating

Mating between genetically individuals that are more closely related by ancestry than in random mating. This system of mating is also known as inbreeding. So it has following effects on a population.

i) It increase homozygosity and decrease the heterozygosity.

ii) Total genetic variability of the population increases rapidly but genetic variability within lines decreases due to random fixation of genes in different families.

iii) Increased the correlation between relatives.

vi) Useful for development of inbred lines.

3. Genetic dis-assortative mating

Mating between genetically dissimilar or less closely related individuals. This system increases the variability, population mean and heterozygosity and reduces homozygosity and correlation. This system is more useful for development of more stable populations.

4. Phenotypic assortative mating

Mating between individuals which are phenotypically more similar than would be expected under random mating. It leads to:

i) This system of mating divides the population into two extreme phenotypes. It increases the variability.

ii) Homozygosity increases.

iii) Prepotency increases.

iv) Useful in isolating extreme phenotypes. It is used in recurrent selection.

5. Phenotypic dis-assortative mating

Mating between phenotypically dissimilar individuals belonging to the same population is called phenotypic dis-assortative mating. The main effects are as under:

i) Maintenance of genetic variability.

ii) Heterozygosity remains unchanged or slightly increases.

iii) Reduction in population variance as it produces intermediate phenotype.

iv) The correlation decreases.

v) Useful in making a population stable.

vi) Used in disruptive mating system.

13

Hybrid and Synthetic Varieties

Hybrids are produced by crossing two genetically diverse parents variety. Hybrid vigour is at its maximum in the F_1 generation and hence highest emphasis is placed on the F_1 hybrids. Hybrid varieties leads to maximum exploition of heterosis as compared to any other varieties. Beal in 1880 recommended the use of F_1 varietal crosses of maize for commercial cultivation.

Features of Hybrid Varieties

1. Hybrids are confined to F_1 generation.
2. Hybrid varieties are highly productive and show vigorous, because heterosis is fully exploited in hybrid varieties.
3. All the plants of a hybrid variety are genetically uniform. The hybrid varieties are heterozygous but homogeneous populations.
4. Hybrids have wider adaptability due to high inherent buffering capacity and gene combination from two divergent parents.
5. Hybrids can be produced in both cross and self-pollinated species depending upon the magnitude of heterosis.
6. Hybrids are more common in cross pollinated species than self-pollinated crops.
7. Hybrids are generally more tolerant to biotic and abiotic stresses than inbreds and pure-line varieties.
8. Heterosis has been commercially exploited in cross pollinated species like maize, pearl-millet, sunflower, castor, onion and cucurbits. It has also been used in some self-pollinated species such as sorghum, cotton, tomato, brinjal, rice etc.

Types of Hybrids

1. Single cross hybrid

When a cross is made between two inbreds (A×B). Single cross exhibit maximum amount of heterosis. Two inbreds were used to produce F_1 hybrids by Shull in

1909. The single cross hybrid were first initiated in maize by Beal. But did not become popular because of poor inbreds.

2. Double cross hybrid

D.F. Jones in 1918 suggested production of double cross hybrids. Double cross hybrids are produced by crossing two single crosses (A × B) × (C × D), involveing four different inbreds. They were superior to single cross hybrids. The first commercial double cross hybrid, Burr Leaming dent was released in 1922. They are more uniform in shape, size and more economicall.

3. Three way cross hybrid

A cross between. single cross and an inbred (A×B) × R. Single cross is used as female and inbred as a male.

4. Top cross hybrid

Top cross is the cross of an inbred with open pollinated variety (Inbred × OP Variety).

5. Poly cross

Progeny of a line developed through, out crossing with other number of selected lines.

6. Population cross

A cross between two open pollinated variety.

Operations Involved in the Production of Hybrid Varieties

The following steps are involved for the production of hybrids

1. Development of inbred lines

- An inbred is a pure line developed by continuous self pollination of plants.
- The source population for isolating superior plants for developing inbreds can be open pollinated varieties, synthetic varieties, composite varieties or any heterozygous population.
- Plants to be selfed are selected on the basis of their superiority like, general vigour, standing ability, plant height, days to maturity, disease resistant and insect resistant etc.
- Self pollinating is achieved by bagging the selected plant at appropriate stage. After self-pollination, the S_1 plants are obtained and growing them row-wise, after repeated selfing of selected plants, S_2 seeds are obtained.

- This process is repeated upto S_5 or S_6 generation. After six generation of selfing, inbred becomes homozygous. The purpose of inbreeding is to fix the desirable characters in homozygous condition in order to maintain them without any genetic change.
- A number of inbred lines are prerequisite for developing hybrids in case of cross pollinated crops. This process is not required in case of self pollinated crops.
- Inbred lines are homozygous genotypes developed by repeated selfing with selection over several generations.

2. Evaluation of Inbreds

When an inbred line is developed, it is crossed with broad base genetic tester and its productiveness in single and double cross combination is tested. The ability of an inbred to transmit desirable performance to its hybrid progenies is referred as its combining ability. GCA is the average performance of an inbred line in a series of cross combination with other inbred lines is termed as general combining ability whereas, SCA is the excessive performance of a cross over and above the excepted performance based on GCA of the parents is known as specific combining ability. Thus GCA is the characteristic of parents and SCA is characteristic of crosses or hybrids. Evaluation of inbreds may be divided into four steps:

i) **Phenotypic evaluation**: Based on the phenotypic performance of the inbreds. It is more effective for characters with highly heritable. Highly performing inbreds are retained and poorly performance are discarded.

ii) **Top cross test**: The inbreds, which are selected on the basis of phenotypic evaluation, are crossed with broad genetic base tester. The top cross progeny are planted in replicated yield trial to see their performance. High yielding progeny are retained and remaining are eliminated.

iii) **Single cross evaluation (A×B):** The remaining inbred lines after top cross test are generally crossed in diallel to test for SCA. Single crosses are evaluated in replicated yield trial. High yielding single cross combinations are identified.

iv) **Double cross performance (A × B) × (C × D):** Jenkins (1934) provided the basis for prediction the average of non parental single crosses. For e.x., with A,B,C,D, inbreds AxB, AxC, AxD, BxC, BxD and CxD single crosses are possible. The predicted yield of double cross (AxB) x (CxD) would be the average of all single crosses leaving AxB and CxD that is parental.

3. Techniques for Hybrid Seed Production

Hybrid seed may be produced in one of the following several ways. The brief description of some of the techniques are as under:

1. Hand emasculation and pollination

The techniques of hand emasculation and pollination can be good method for producing hybrid seeds or for curtained choice flowers, vegetables or fruits where single pollination produces large number of seeds and the seed is needed for planting only on a limited scale. It has been successfully employed in hybrid seed production of crops like tomato, brinjal, watermelon, cucurbits, squashes etc.. In case of maize detasseling is the process of manual or mechanical removal of immature tassel before they start shedding their pollen foe preparation of seed parent in the hybrid seed production. However, large scale seed production sometimes becomes handicapped because of certain problems. To cope up with such of the problem, male sterility and self-incompatibility offer the means for genetic emasculation. Pollen dispersal is often satisfactory in most cross-pollinated species since it is their natural mode of reproduction. But in self-pollinated species, satisfactory pollen dispersal is often the limiting factor in hybrid seed production.

2. Cytoplasmic – genetic male sterility

The cytoplamic male sterility is a type of male sterility which is governed by the interaction of nucleus and cytoplasm. The male sterile plants have sterile cytoplasm while the fertile plant has fertile cytoplasm. The progenies of male sterile plants are usually male sterile e.i., it is transmitted by the maternal plant. This type of male sterility is maintained by crossing it with maintainer line containing male fertile cytoplasm. This system is the most widely used in hybrid seed production in crops like, maize, pearlmillet, jowar, onion, sugarbeets, cotton, rice, etc. The system is based on a cytoplasm that produces male sterility and on a gene that restore fertility in the presence of the male sterile cytoplasm. The use of this system in hybrid seed production is outlined below:

i) Production of single cross hybrid varieties

Male sterile line is used as female parent and the male parent as a restorer line for the production of a single cross. The seed set on the female parent (male sterile line) is the hybrid seed, while seed produced on the male parent is selfed seed. The resulting hybrid is male fertile as it has received the restorer gene from the male parent. Generally, two rows of the male fertile inbreds are planted after every four rows of the male sterile. The ratio of male and female inbreds is kept 2:4. for the production of single cross hybrid varieties of maize.

ii) Production of double cross hybrid varieties

Double cross hybrid varieties are developed by crossing two single crosses (AXB) (CxD), one is male sterile and the other is male fertile. The male sterile single cross is produced by crossing a cytoplasmic male sterile line with a non-restorer male fertile line. The male fertile single cross may be produced in one of the two ways. First, a cytoplasmic male sterile line is crossed with a restorer line. The double cross in this case has both male sterile and male fertile plants in the ratio 1:1. In the second method, two restorer lines are crossed together; one of the restorer lines serves as female and is detasselled manually. All the plants in double cross would be male fertile.

iii) Genetic male sterility

Genetic male sterility is a type of male sterility in which the expression of the character is governed by genetic factor. In a number of crop plants, the genetic male sterility is governed by homozygous recessive genes (msms). Such type of MS is maintained by crossing male sterile plants with heterozygous fertile plants. Half of the progeny would be sterile and half would be heterozygous fertile. In India, it is being used for hybrid seed production of arhar.

iv) Cytoplasmic male sterility

This type of male sterility is determined by the cytoplasm. Since the cytoplasm of a zygote comes primarily from egg cell, the progeny of such male sterile plants would always be male sterile. This type of male sterility may be utilized for producing hybrid seed in certain ornamental species, or in species where a vegetative part is of economic value. But in those crop plants where seed is the economic part, it is of no use because the hybrid progeny would be male sterile.

v) Self incompatibility

Self incompatibility refers to the inability of a plant to set seed when self pollinated with normal functional pollen grain and ovules even though it can set seed when cross pollinated. Two self-incompatible but cross compatible lines are planted in alternate rows, the seed produced by both the lines would be hybrid seed.

Merits of Hybrid Varieties

1. Hybrid varieties are of higher yield potential than synthetics, composites and open pollinated varieties.
2. Hybrid varieties are more uniform and attractive as compared to synthetics, composites and open pollinated varieties.

3. Hybrid varieties can be produced in both self and cross pollinated species, whereas synthetics and composites are relevant to cross pollinated crops.
4. Hybrid can be reconstitute with the same genotype, which is not possible in case of composite varieties and open pollinated varieties.

Demerits of Hybrid Varieties

1. Fresh seed has to be produced every year, because in F_2 the hybrid produces various types due to segregation and recombination.
2. Farmer has to purchase new seed every year.
3. Farmers need more money to purchase hybrid seed.
4. The seed of hybrid is costlier than synthetics, composites and open pollinated varieties.
5. Production of hybrids requires more input viz; fertilizer, water, plant protection, etc. to exploit their full potential.
6. Cultivation of hybrid requires more technical skill.

Synthetic Varieties

Synthetic varieties have been of great value in the breeding for those cross pollinated crops, where pollination control is difficult. For example Alfalfa, cloves, forage crop species etc. Even in maize improvement of synthetic varieties are becoming increasingly important and in India there had been a considerable emphasis on synthetic development. The possibilities of commercial utilization of synthetic varieties in maize were first suggested by Hayas and Garber in 1919. Synthetic variety is developed by inter-mating in all possible combinations, a number of inbred lines with good general combining ability and mixing their seeds in equal quantity are known as synthetic variety. Synthetic variety is maintained by open pollination in isolation, some degree of selfing occurs resulting in fixation of some genes. It is relevant to cross pollinated species. The basic concept in the development of synthetic variety is exploitation of heterosis. Synthetics variety exploited only partially heterosis, because some degree of inbreeding takes place due to open pollination in later generations. Materials used for development of synthetic varieties may be inbreds, clones, open pollinated varieties and the end products of recurrent selection, which are already tested for GCA. Synthetic varieties developed in India are, sugarbeet-plant synthetic-3, Cauliflower- synthetic-3, bajara-ICMS-7703.

Features of Synthetic Variety

1. Heterogeneous.
2. Clones, inbreds or. open pollinated variety can be used for development of synthetic variety.
3. They are relevant to cross pollinated species.
4. They are maintained by open pollination.
5. Synthetic variety can be reconstituted.
6. More adaptive to varying growing conditions as compared to hybrids.
7. Less uniform and attractive as compared to hybrids.
8. Show partial amount of heterosis as compared to open pollinated variety.
9. Better disease resistance.

Development of Synthetic Varieties

The various operations involved in the production of synthetic varieties:

1. Evaluation of lines for GCA

Parents of synthetic variety may be inbred lines, clones, open pollinated variety or short term inbred lines. GCA of parental lines is generally estimated by top cross or polycross test. Evaluation of GCA is done because in synthetic variety, that portion of heterosis is exploited, which is produced by GCA. The lines having high GCA are selected as parents of synthetic variety.

2. Production of synthetic variety

Two steps are used in production of synthetic variety:

i) Equal amount of seed from all the parental lines (syno) are mixed and planted in isolation. Open-pollination is allowed and produce crosses in all combinations. Generally, 5-8 good general combining inbred lines are used to constitute a synthetic variety. The seed from this population is harvested in bulk; the population raised from this seed is the Syn_1 generation.

ii) The parental lines are planted in a crossing block, and all possible crosses among the selected lines are made in isolation. Equal amount of seed from each cross is composited to produce the synthetic variety. The population derived from this composited seed is known as synthetic one generation.

3. Multiplication of synthetic variety

i) When synthetic variety is synthesized. It is multiplied in isolation for one or more generation, before its distribution for cultivation. This is done to produce commercial quantities and is a common practice in most of the crops.

ii) A synthetic variety can safely be grown for a period of 4-5 years without reduction in yield potential. After a period of five years, it would be desirable to reconstitute the synthetic variety. Farmers can use their own seeds upto five years. They are maintained by open pollination. They exploit more of additive variance.

Systematic Representation for Production of Synthetic Varieties

First year: Selection of good inbred lines, crossing them with a common tester, harvesting top crossed seed separately.

Second year: Evaluation of top crosses in replicated yield trials using standard hybrid or open pollinated variety as check. Identification and selection of good general combining inbred lines the basis of top cross performance.

Third year: Making all possible single crosses among the selected inbred lines for good GCA and harvesting crossed seed of single crosses separately.

Fourth and fifth year: Mixing seed of all single crosses in equal quantity and seed multiplication by open pollination in isolation for one or two generations.

Merits of Synthetic Variety

1. Cost is less as compared to hybrids.
2. Farmer can use synthetic varieties for 3 to 4 years, which is not possible in hybrids.
3. Synthetics are more stable over years and environments.
4. More techniques are not required for seed production in synthetics, as it is more in hybrids.

Demerits of Synthetic Variety

1. Yield potential of synthetic is less as compared to hybrids because synthetics exploit only GCA while hybrids exploit both GCA and SCA.
2. The performance may not be good when lines having low GCA.

Composite Varieties

A variety which is developed by mixing the seed of various genotypes in equal quantity, which are similar in maturity, plant height, seed size, colour etc. is called composite variety. The lines used to produce a composite variety are rarely tested for combining ability. Several composite varieties have been developed in maize such as, Jawahar, Sona, Vikas, Vijay, Vikaram, Kiran, Protina, Kishan, Ratan, Kanchana, Amber, Shakti, Compositae 1 in *B. campestris,* Parbhani Sampada and Mandor Bajara in pearlmillet.

Features of Composite Varieties

1. Composite varieties are developed by mixing the seed of various genotypes which are similar in maturity, height, seed size, colour, etc.
2. The variety is maintained by open pollination.
3. Farmers can use their own seed for 3-4 years.
4. Composite varieties utilize heterosis.
5. Highly adaptable to environmental fluctuations.
6. Since composites are developed from heterozygous genotypes, it is not possible to reconstitute the composite varieties, because in segregating populations gene frequency changes with time.
7. Composite varieties consist of several homozygotes and heterozygotes. constituting a heterogenous population.

Development of a Composite Variety

Several steps are used for development of composite variety:

i) Select diverse varieties or line.

ii) Make all possible crosses among them and harvesting crossed seed separately.

iii) Select heterotic crosses and advanced them to F_2, F_3 and F_4 generation.

iv) Evaluate the segregating generations and compare with best hybrid.

v) Select high yielding crosses.

vi) Mix seeds of superior crosses in equal quantity to constitute composite variety.

vii) Best one released as a variety.

14

Breeding Methods for Vegetatively Propagated Crops

Reproduction that does not involve the union of male and female gametes is known as asexual reproduction. Reproduction in which sexual organs or related structures take part but fertilization does not occur, or crops which are propagated asexually or by vegetative means are known as asexually propagated crops. Most of the fruit plants are propagated asexually which consist of large number of clones. Example: sugarcane, potato, sweet potato, banana, mango, citrus, pears, peaches, litchi, crops that propagates by asexual means. The procedure of selection used for asexually propagated crops is known as clonal selection, since the selected plants are used to produce new clones or superior clones can be isolated from local variety, introduced variety or inter crossed population.

Breeding of Clonal Crops

First Year

i) Desirable plants are selected from a mixed variable population.

ii) A vigorous selection should be done for simply inherited characters with high heritability.

iii) Poor plants should be eliminated.

Second Year

i) Clones from the selected plants are grown separately.

ii) Clones are selected on visual observation.

iii) Finally, fifty to one hundred clones may be selected on the basis of clonal characteristics.

iv) Inferior clones are eliminated.

Third Year

i) A replicated preliminary yield trial is conducted along with suitable check for comparison.

ii) Few superior performing clones with desirable characteristics are selected for multi location trials.

Fourth to Sixth Years

i) A replicated yield trial is conducted at several locations along with a suitable check to see the performance of newly developed clones.

ii) The clone superior to the check variety in one or more characteristics is identified for release as a new variety.

Seventh Year

The superior clone is multiplied and released as a new variety.

Hybridization in Clonal Crops

Clonal crops are generally improved by mating or crossing of two or more desirable clones, followed by selection in the F_1 progeny and in the subsequent clonal generations. The improvement through hybridization involved three steps-

1. Selection of parents

i) The selection of parents to be employed in hybridization is the most crucial step, since the value of F_1 progeny would depend upon the parents used.

ii) Selection of parents generally based on their known performance both as varieties and as parents in hybridization programme.

iii) The performance of a clone in hybridization programme depends on its prepotency and general combining ability.

iv) The selected parents are used to produce single crosses involving two parents.

2. Selection among F_1 Families

i) Large number of crosses have to be made in order to ensure that at least some of the crosses would produce outstanding progeny in F_1.

ii) The F_1 families or population are observed visually.

iii) Inferior families are eliminated.

iv) Promising families with outstanding individuals are then grown at a much larger scale for selection.

3. Selection within F_1 Families

The selection procedure within F_1 families is essentially the same as that in the case of clonal selection.

Procedure

First Year

Clones to be used as parents are grown and crosses are made to produce F_1 progeny.

Second Year

Large number of seedlings obtained through hybridization are space planted. Few hundreds to few thousand desirable plants are selected. Undesirable plants are discarded.

Third Year

Individual clones are grown separately. Upto 200 superior clones may be selected for preliminary yield trial. Poor clones are eliminated.

Fourth Year

A replicated preliminary yield trial is conducted along with suitable check. Few outstanding clones are selected for multi locations trials.

Fifth to Seventh Year

Multilocation yield trials are conducted at several locations with standard varieties as check. Clones superior to the check are identified for release as new varieties.

Eighth Year

The clones released as varieties are multiplied and distributed among farmers.

Characteristics of Asexually Propagated Crops

1. Most of these plants are perennial and annual like sugarcane, potato, sweet potato.
2. Many of the crop shows reduced flowering and seed set.
3. Many crop varieties do not flower at all.
3. They are invariably cross-pollinated.
4. They are highly heterozygous and show severe inbreeding depression.
5. Many species are interspecific hybrid. Example: Banana, sugarcane.
6. Many species shows wider adaptation.
7. These crops consist of a large number of clones.

15

Mutation Breeding

Mutagenesis is the process whereby sudden heritable changes occur in the genetic information of an organism not caused by genetic segregation or genetic recombination, but induced by chemical, physical or biological agents.

Mutations were first identified by Hugo de Vries in the late nineteenth century, while experimenting on the 'rediscovery' of Mendel's laws of inheritance. He coined the term 'mutation'. Radiation-induced mutations as a tool for generating novel genetic variability in plants, advanced as a field after the discovery of the mutagenic action of X-rays demonstrated in maize, barley and wheat by Stadler. The discovery of X-rays by Rontgen in 1895 led to the application of X-rays for inducing mutations in fruit fly by Muller (1927). Mutation is of two types:

1. Spontaneus mutations

Mutations do occur in nature by natural phenomena like cosmic and ultra violet rays. But the rate of these spontaneous mutations is very low i.e. 10^{-6} or one out of million for an individual gene.

2. Induced mutations

Mutations induced artificially by certain physical and chemical mutagens are referred as induced mutation.

Features of Mutation

The main features of mutations are summarized below:

1. Mutations may occur at any stage in the development of an organism.
2. Mutations are mostly harmful to the organism but a very small portion of them may be useful (0.1%) for crop improvement.
3. Mutations are usually recessive but dominant mutations also take place.
4. Mutations may take place either in somatic cell or germinal cells.
5. Mutations may be large or small.

6. Mutations may occur in any gene.
7. Mutations may arise due to natural forces or may be induced artificially.
8. Mutations are recurrent that same mutation may occur repeatedly.
9. Mutations occur in both forward and reverse directions.
10. Most mutant alleles are pleiotropic.

Classification of Mutation

1. On Phenotypic Effect

i) **Macro mutations:** These are large mutations; produce a large recognizable phenotype effect on individual plants. These are governed by oligogenic gene and can be easily selected in the M_2 generation.

ii) **Micro mutations:** This type of mutation produce a small phenotypic effect that can be identified only on the basis of a population. These are polygenic in nature and selection for such mutations can be delayed till M_3 or later generations.

2. Based on their Detection

i) **Morphological mutations**: Involve change in external form including colour, shape, size etc. e.g. Kernel colour in corn and dwarfism in pea.

ii) **Biochemical mutations**: They are identified by a deficiency so that the defect can be overcome by supplying the nutrient or any other chemical compound for which the mutant is deficient. Such mutations have been studied in prokaryotes like bacteria and fungi.

3. Based on Survival

i) **Lethal mutations**: Kill each individual that carry them. These are perhaps easiest to score e.g. albino individuals resulting from chlorophyll deficiency are lethal.

ii) **Sublethal mutation**: Kill more than 50% individuals.

iii) **Subvital mutation**: Death less than 50% individuals.

iv) **Vital mutation**: Do not kill any individual that carry them. They are useful for plant breeding point of view.

5. Basis of cell origin

i) **Somatic mutations**: Mutations that occur in the body cells or somatic cells are termed as somatic mutations. The changes that take place by somatic mutations are not transmitted to the next generation.

ii) **Germinal mutations**: Germinal mutations occur in reproductive cells or germ cells that produce gametes. Only these mutations are transmitted to the next generation. So, these are also known as heritable mutations and are significant from evolutionary point of view.

6. Molecular Basis

i) **Base substitution**: In a substitution mutation a nitrogenous base of triplet codon of DNA is replaced by another nitrogenous base that changes the codon. Single base substitution is also known as point mutations. The substitution may be of the following two types:

a) **Transition**: When a purine is replaced by another purine (A—G) or pyrimidine is replaced by another pyrimidine (C—T).

b) **Transversion** : The exchange of a purine (adenine and guanine) by a pyrimidine (thymine and cytosine) or a pyrimidine by a purine is called as transversion. Thus, eight different types of possible transversions are: G C or A C G T A TC G or TG C A T A.

ii) **Base insertions and deletions** : Addition of one or more extra nucleotide is called insertion or removal of one or more bases is called deletion from the DNA of a gene. The number can range from one to thousands. A point mutation may consist of addition or deletion of a base. This kind of mutation is known as frame shift mutation.

7. Resistant Mutations

They are identified by their ability to grow in the presence of an antibiotic or pathogen to which the control is susceptible.

8. Conditional Mutations

Mutations which allow the mutant phenotype to be expressed only under specific conditions, (e.g. high temperature) called restrictive conditions, under normal or permissive conditions, the mutant express normal phenotype. These have been used to study cell cycle.

Mutagens and their Mode of Action

The agents available for induction of mutations i.e. mutagens can be categorized into following two classes:

1. **Physical mutagens:** X-rays, Gamma rays, UV radiations, Alpha-particles, Neutrons and Particles from accelerations.

2. **Chemical mutagens:** Base analogues, Antibiotics, Alkylating agents, Acridines, Azides, Hydroxylamine and Nitrous acid.

1. Physical Mutagens

The physical mutagens includes ionizing and non–ionizing radiations. Ionizing radiation can penetrate in to the cells and create ions in the cell contents. Ionization radiations are of two types - particulate and non–particulate. Particulate are high–potential, fast–moving charged atomic particles produced by radioactive decay.

i) X-rays

They are produced by X–rays machine. Roentgen 1895, first time invented the X-rays. They break the chromosome and produce all types of muations. X-rays were first used by Muller in 1927 to induce mutations in Drosophilla.

ii) Gamma radiation

They are obtained from radioactive cobalt (^{60}Co) and is widely used. It has high penetrating potential and is hazardous. However, it can be used for irradiating whole plants and delicate materials, such as pollen grains. Various mutants have been developed through gamma radiation.The mutagenic effect results mostly from DNA double-strand breaks. The mutants show higher potential for improving plant architecture leading to better crop improvement and are used as a complementary tool in plant breeding. Gamma rays have a shorter wave length and therefore, possess more energy than protons and X-rays, which gives them ability to penetrate deeper into the tissue.

iii) Beta radiation

Beta particles consist of light particles, each carrying a single negative electrical charge. These negative beta particles are identical to the electrons. They are sparsely ionizing and more penetrating than alpha particles. They are generated from 3_H, 32_P, 35_S etc.

iv) Alpha particles

Alpha particles are composed of two protons and two neutrons. They have double positive charge. They are densely penetrating, but less penetrating in comparison to beta rays and neutrons.They are emitted from isotopes of heavier elements.

v) Fast and thermal neutrons

They may be obtained from atomic reactor 235_U. They are neutral and are not stopped by negative or positive charge tissue. Neutrons are hazardous and hence have less penetrating abilities, but they are known to cause serious damage to the chromosomes. They are best used for materials, such as dry seeds. Various forms of neutrons were also studied extensively for their use in mutagenesis. Though it has been proved to be an effective mutagen, particularly for producing large DNA fragment deletions, the application of neutrons in induced mutagenesis is limited.

vi) UV radiation

The UV radiations possess limited tissue penetrating ability as a result of which their use is restricted to pollen grains. They are produced from mercury vapour lamps. However; UV radiation is fairly good for irradiation of cultivated plant cells/protoplasts. The mutagenic effect of ultraviolet light was discovered by Altenbung through irradiation of the polar cap cells of fruit fly eggs. The mutagenic potential of these rays have since been confirmed in many organisms. In those organisms, germ tissue could be easily exposed to the low-penetrating ultraviolet light which resulted in covalent dimerization of adjacent pyrimidine. Emission of UV light (250–290 nm) has a modest capacity to infiltrate tissues as compared with ionizing radiation.

In general, ionizing radiations such as X-rays, gamma rays are preferred because of their easy application, good penetration, reproducibility, high mutation frequency and less disposal problems. The radiation dose is determined by the intensity of radiations and length of explosive. It is expressed in Roentgen (R) units, which are a measure of the number of ionizations that occur. In the mutation breeding experiments, irradiation dose is generally expressed as kR or Gray (Gy) where 1 Gy=100 rad and 1 kR = 10 Gy. The unit of absorbed dose is rad (radiation absorbed dose) where 1 rad= 100 erg/g= 10-2 joule/kg and it is expressed as rad per second or per minute or per hour.

2. Chemical Mutagens

The effect of chemical mutagens on plant materials is generally considered milder. The ratio of mutational to undesirable modifications is generally higher for chemical mutagens than for physical. However, chemical mutagens are generally carcinogenic. Despite the large number of mutagenic compounds, only a small number has been used in plants. Among them, only a very restricted group of alkylating agents has found large application in plant mutation breeding.

i) Alkylating agents

Addition of an alkyl group to the hydrogen bonding oxygen of guanine (N_7 position) and adenine (at N_3 position) residues of DNA is done by alkylating agents. As a result of alkylation, possibility of ionization is increased with the introduction of pairing errors. Hydrolysis of linkage of base-sugar occurs resulting in gap in one chain. This phenomenon of loss of alkylated base from the DNA molecule (by breakage of bond joining the nitrogen of purine and deoxyribose) is called depurination. Depurination is not always mutagenic. The gap created by loss of a purine can effectively be repaired. They can be found among a large array of classes of compounds, including sulphur mustards, nitrogen mustards, ethyleneimines, ethyleneimides, alkyl methanesulphonates, alkylnitrosoureas, alkylnitrosoamines, alkylnitrosoamides, alkyl halides, alkyl sulphates, alkyl phosphates, chloroethyl sulphides, chloroethylamines, etc. One of the most effective chemical mutagenic groups is the group of alkylating agents (these react with the DNA by alkylating the phosphate groups as well as the purines and pyrimidines).

ii) Base analogues

A base analogue is a chemical compound similar to one of the four bases of DNA. It can be incorporated into a growing polynucleotide chain when normal process of replication occurs.' These compounds have base pairing properties different from the bases. They replace the bases and cause stable mutation. A very common and widely used base analogue is 5-bromouracil (5-BU) which is an analogue of thymine. The 5-BU functions like thymine and pairs with adenine. The 5-BU undergoes tautomeric shift from keto form to enol form caused by bromine atom. The enol form can exist for a long time for 5-BU than for thymine if 5-BU replaces a thymine, it generates a guanine during replication which in turn specifies cytosine causing G: C pair. During the replication, keto form of 5-BU substitutes for T and the replication of an initial AT pair becomes an A: BU pair. The rare enol form of 5-BU that pairs with G is the first mutagenic step of replication. In the next round of replication G pairs with C. Thus, the transition is completed from AT : GC pair.

iii) Intercalating agents

There are certain dyes such as acridine orange, proflavine (Ethidium bromide) and acriflavin which are three ringed molecules of similar dimensions as those of purine pyrimidine pairs. In aqueous solution these dyes can insert themselves in DNA (i.e. intercalate the DNA) between the bases in adjacent pairs by a process called intercalation. Therefore, the dyes are called intercalating agents. The acridines are planer molecules which can be intercalated between the base pairs of DNA; distort the DNA and results deletion or insertion after replication of DNA molecule. Due to deletion or insertion of intercalating agents, there occur frameshift mutations.

Procedure of Mutation Breeding

The first step in plant breeding is to identify suitable genotypes containing the desired genes among existing varieties, or to create one if it is not found in nature. In nature, variation occurs mainly as a result of mutations and without it, plant breeding would be impossible. In this context, the major aim in mutation-based breeding is to develop and improve well-adapted plant varieties by modifying one or two major traits to increase their productivity or quality. Both physical and chemical mutagenesis is used in inducing mutations in seeds and other planting materials. Then, selection for agronomic traits is done in the first generation, whereby most mutant lines may be discarded. The agronomic traits are confirmed in the second and third generations through evident phenotypic stability, while other evaluations are carried out in the subsequent generations. Finally, only the mutant lines with desirable traits are selected as a new variety or as a parent line for cross breeding. The following scheme can be used:

i) About 500-600 seeds are treated with an appropriate mutagen. The plants raised from the treated seeds constitute M_1 generation. The M_1 is densely planted and seeds from individual plants are harvested separately. The M_1 plants should be observed critically to look for dominant mutations, if any.

ii) The M_2 generation is raised from seeds harvested from M_1 single plants in plant to progeny rows. Single plants from progeny rows suspected to contain mutant alleles are harvested separately. The recessive mutations are expressed in M_2. To achieve the desired target of mutation breeding programme, the size of M_2 generation should be large enough. For polygenic traits in M_2, the vigorous plants are selected and harvested separately to raise individual plant progenies in M_3.

iii) The selected M_2 plants are grown in progeny rows to raise M_3 generation. In M_3 generation the homogeneous mutant progenies are harvested as bulks, whereas the heterogeneous progenies are further subjected to selection of mutant plants, if any, which can be grown to raise M_4 homogeneous progenies.

iv) In M_4 generation, the uniform mutant progenies are further evaluated in replicated trial. The segregating progenies, if any, must be rejected.

v) In $M_{5,}$ the superior mutant progenies are evaluated in preliminary yield trial along with a suitable check. The promising stable mutant progenies/lines are further tested.

vi) Promising genotype is identified for release as a variety.

Advantages of Mutation Breeding

It is cost effective, quick, proven and robust. In addition, mutation breeding is transferrable, ubiquitously applicable, non-hazardous and environmentally friendly. There are more than 3346 mutant varieties officially released for commercial use in more than 228 plant species from more than 73 countries. Mutation breeding has contributed significantly by giving nearly 34 varieties of different pulse crops for commercial cultivation whereas, approximately 97 varieties have been developed at world level.

Achievements

Pusa-408 (Ajay), Pusa-413 (Atul), Pusa-417 (Girnar), Sadhbhawana chickpea, Surya (WCG-2) chick pea, Vallabh kallar channa 1 (WCG-3) in gram. Pant moong-2, TAP-7, TARM-1 in moongbean, TAU 1,2,4 in urd bean, RMO, 40, 257, 435, 2251 in mothbean, Padmini, Mohan, Jagannath, Prabhani and Shyama in rice Sharbati sonara, NP 111 and Pusa herma in wheat, TG 17, TG 1, TG 18, TG 19, TG 7 in groundnut, TA 7-5, TAT-7, CO 3 and CO 5 in Pigeonpea, Knanti, RLM 198, RLM-514 and RLM 619 in mustard and R 1-1, Niharika in isabgol.

16

Polyploidy Breeding

Polyploidy, the condition in which a normal diploid cell or organism acquires one or more additional sets of chromosomes. In other words, the polyploid cell or organism has three or more times the haploid chromosome number. Polyploidy arises as the result of total nondisjunction of chromosomes during mitosis or meiosis. The somatic chromosome number of any species, whether diploid or polyploidy is designated as 2n. The somatic cells of plants are diploid that they posses two complete set of chromosomes, one received from seed parent and another from pollen parent. The gametic chromosome number is said to be haploid and denoted by n. Haploid is defined as cell with gametic chromosome number. While monoploid has the basic chromosome number and represented by x.

Classification of Polyploidy

Polyploids are classified in to two groups, euploidy and aneuploidy. Euploidy have chromosome number which is an exact multiple of basic number. Euploidy includes monoploids, diploids and polyploids. In aneuplod individual with other than exact multiple of basic chromosome number or change in chromosome number which posses one or few chromosomes of the genome.

1. Euploidy

It refers to the situation in which an organism has one complete set of chromosomes or an integral multiple of a set. These are of the following types:

I. Monoploid

The number of chromosomes in a basic set is called monoploid number (x). (i) Monoploidy can arise from spontaneous development of an unfertilized egg (parthenogenesis); (ii) Lethal; (iii) Monoploids are usually sterile because of problems with meiosis; no pairing partner for each chromosome. (iv) Monoploids are important in plant breeding. (iv) Monoploid can be haploid but all haploid can not be monoploids.

II. Haploid

Haploid number (n) refers to number of chromosomes in gametes. For most of the diploid organisms, the haploid and monoploid number is the same. However, this is not always the case; wheat is a hexaploid containing six sets of similar but non–identical chromosomes with 7 chromosomes per set (x=7 and 6x = 42). The gametes however, have 21 chromosomes (n=21 and 2n=42). The 'normal' situations are 1x = haploid or 2x = diploid. (i) Haploid may be originated spontaneously due to parthenogenesis. (ii)Haploids are used for development of pure lines, inbred lines, aneuploids and disease resistance. (iii) They may be artificially produced by induction of X-rays, delayed pollination, temperature shocks, colchicine treatment, distant hybridization, anther and pollen culture.

III. Polyploids

The presence of more than two genomes in an individual is called polyploids. About one third species of flowering plants are polyploid. It has been reported that polyploidy are found in about seventy percent of wild species of grassy family. There are mainly two kinds of polyploids:

i) **Autopolyploids:** They are polyploids with multiple chromosome sets derived from a single species. They arise spontaneously as in, naturally occurring tetraploid species like potato. Others might develop following fusion of 2n gametes (unreduced gametes). Bananas and apples can be found as autopolyploids. Autopolyploid plants typically display polysomic inheritance, and are therefore often infertile and propagated clonally. Autopolyploid includes triplods (3x), tetraploid (4x), pentaploid (5x), hexaploid (6x) etc. Example, potato, banana, coffee, peanut, alfalfa. peppermint and sweet potato.

Applications of Autopolploidy

Autopolyploids can arise from spontaneous aberrant meiosis in nature to give a 2n gamete that fuses with a normal n gamete to give a 3x zygote or from tetraploid (4x) and diploid (2x) cross. These are usually sterile because the frequency of getting exactly 1 set or exactly 2 sets of all the chromosomes is to migrate to one pole. Gametes that contain intermediate chromosome between n and 2n exhibit genome imbalance because there are different ratios of genes depending upon 1 or 2 copies of each chromosome. Triplods are developed by crossing tetraploid and diploid. They are used for the production of seedless watermelon, apple,sugarcane and sugarbeets. Sugarbeet that give rise to larger roots and more sugar per unit area. They are sterile, but propagate vegetatively.

ii) **Alloploidy** (multiple sets from different species) : They may also result from complete chromosome sets from two or more species. e.g., Triticale and Raphano brassica. Allotetraploids (4 sets from different species: 2 set from species 1 and 2 sets from another species). Chromosomes are homeo-logous (similar but not completely homologous). They can be developed by inter specific crosses and fertility is restored by doubling the chromosome number with colchicine. Allotetraploidy is important in the evolution of plants (cotton, wheat, brassica, tobacco etc.).

Applications of Allopolyploidy

Alloploidy sometimes leads to the creation of new crop species. Triticale is the best example, which is allohexaploid between wheat and rye. It combines desirable character of both the species i.e., grain quality of wheat and hardiness of rye. Presently cultivated strawberry originated from a cross between North and South American species in the middle of seventeenth century in Europe. Loganberry was developed from a cross between raspberry and blackberry in 1880 in California, USA.When the desirable character is not found within the species, it is transferred from the related species, called interspecific gene transfer. Interspecific gene transfer is done in two ways, viz., by alien addition and alien substitution. In case of alien addition, one chromosome of wild species is added to the normal complement of a cultivated species. In case of alien substitution, one pair of chromosome is substituted in cultivated species with those of wild donor species. Such type of gene transfer has been achieved in crops like wheat, tobacco, cotton and oats. In cotton, lint strength has been transferred from *G. thurberi to G. hirsutum*. However, in such gene transfer several undesirable characters (genes) are transferred along with desirable ones. In tobacco, mosaic resistance has been transferred from *N. glutinosa to N. tabacum* through alien addition.

Sometimes direct cross between two species is not possible due to sterility in F_1. In such case, first an amphidiploid is made between such species and then amphidiploid is crossed with the recipient species. Such types of bridging crosses have been made for transfer of genes from wild species particularly in crops like tobacco and cotton.

Besides these, alloploids have some other applications. For example, heterotic effects can be conserved more easily in allotetraploids than in diploids. Moreover, interspecific hybrids can be made fertile by artificial doubling of chromosomes.

1. Wheat

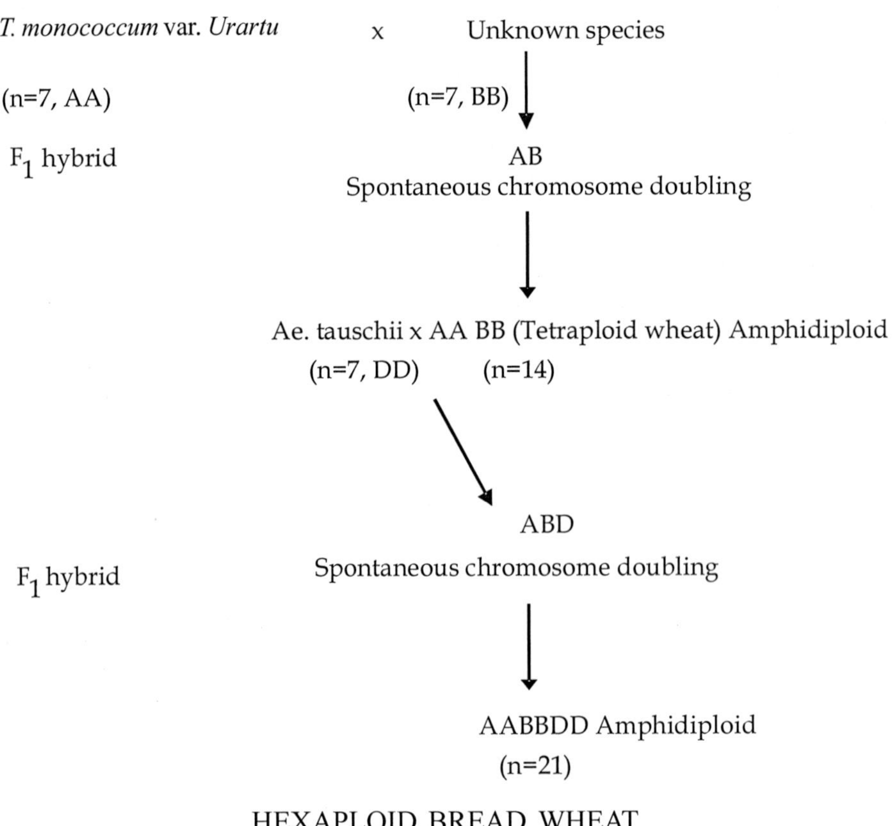

HEXAPLOID BREAD WHEAT

2. Cotton

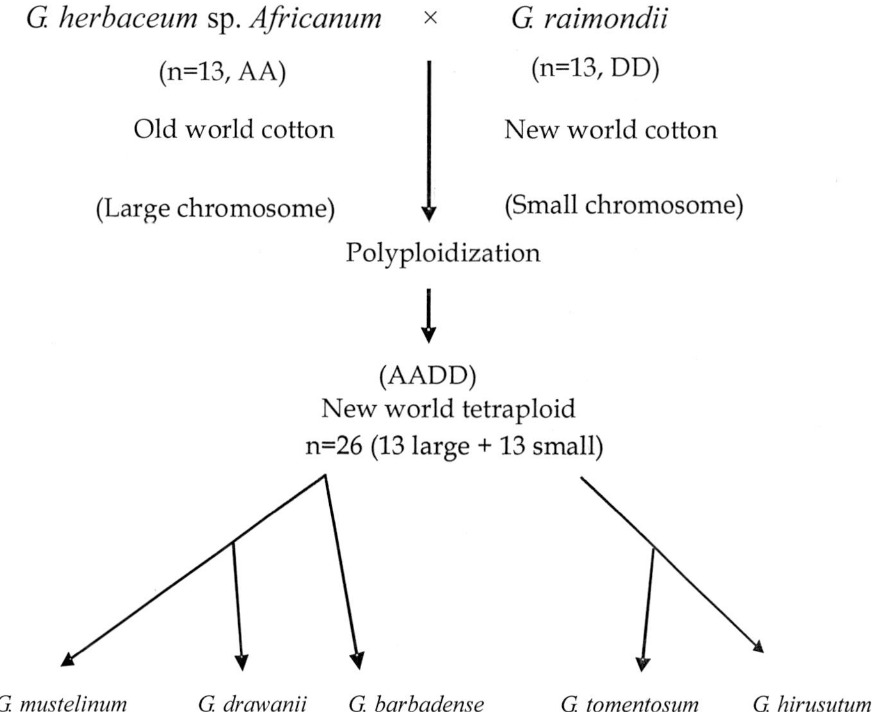

3. Mustard

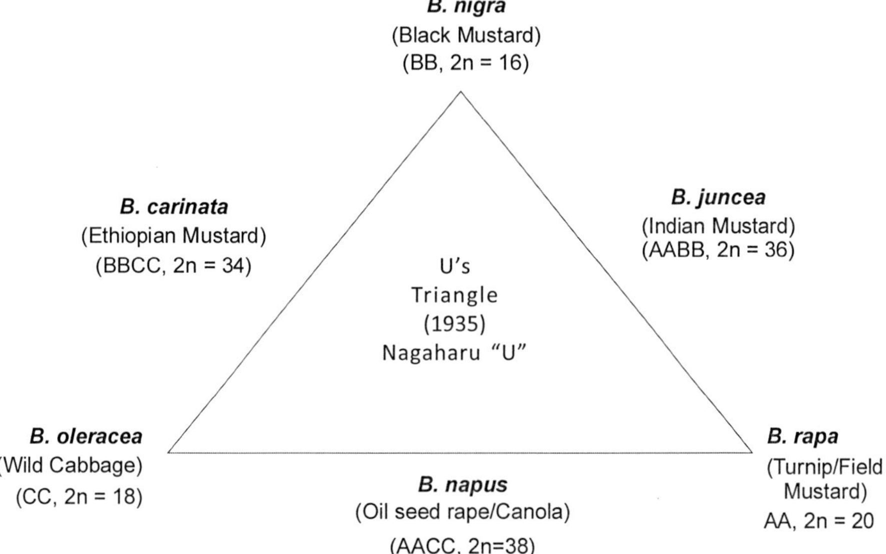

2. Aneuploidy

The situation in which one or several chromosomes are lost or added to the normal set of chromosomes is called aneuploidy. It is caused by non–disjunction during meiosis to generate n+1 or n-1 gametes which when fertilized with n gamete generate 2n+1 or 2n–1 zygotes.

a) Nullsomics: Missing of a single pair of homologous chromosomes (2n–2). Viable only in some polyploid species.

b) Monosomics: Missing of one complete chromosome (2n-1). Can be viable in some plants, in fact, monosomic banks can be used to map recessive alleles to the chromosomes.

c) Trisomy: They have one extra chromosome of a certain chromo-some (2n+1). When extra chromosome is identical to its homologues, it is called primary trisomics. Extra chromosome should be an isochromosome and should be the product of a tranlocation.

d) Tetrasomy: Having one pair of chromosome extra (2n+2). All the 21 tetrasomics are available in wheat.

17

Abiotic Stresses (Drought)

Abiotic stresses include salinity, drought, flood, extremes in temperature, heavy metals, radiation etc. They are the foremost factors that causes the loss of major crop plants worldwide. This situation is going to be more rigorous due to increasing desertification of world's terrestrial area, increasing salinization of soil and water, shortage of water resources and environmental pollution.

The inadequacy of water availability, including precipitation and soil moisture storage capacity, in quantity and distribution during the life cycle of a crop to restrict the expression of its full genetic yield potential". Generally drought stress occurs when the available water in the soil is reduced and atmospheric conditions cause continuous loss of water by transpiration or evaporation. Drought stress tolerance is seen in almost all plants but its extent varies from species to species and even within species.

Effect of Drought Stress

1. Reduction of water content.
2. Diminished leaf water potential and turgor loss.
3. Closure of stomata.
4. Decrease in cell enlargement and growth.
5. Arrest of photosynthesis.
6. Disturbance of metabolism and finally the death of plant.
7. Water stress inhibits cell enlargement.
8. Reduces plant growth.

Mechanism of Drought

1. Drought escape

The condition where a drought susceptible variety performs well in a drought conditions simply by avoiding the period of drought. Generally, drought takes place in the late maturing crop. It is most common, in plants which are grown in arid conditions. Early maturity is an important attribute of drought escape, and is suitable for crops subjected to late-season drought stress. Early varieties generally have lower leaf area index, lower evapo-transpirational and lower yield potential. Example: Mothebean variety like, RMO-60, 257, 435, 2251, matures in 65 days, before water deficit develops.

2. Dehydration avoidance

It is the ability of a plant to retain a relatively high level of hydration under conditions of soil or atmospheric water stress. The various physiological, biochemical and metabolic process of plants that are involved in the growth and yield are protected from water stress. The common measures of dehydration aviodance are tissue water status as expressed by water turgor potential under condition of water stress. This can be achieved by abscisic acid, cuticular wax, leaf characteristics, reduced transpiration, osmotic adjustment and increase water uptake.

3. Drought tolerance

The ability of plants to continue metabolizing at low leaf water potential and to maintain growth despite dehydration of the tissue or to recover after release from stress conditions. Translocation is one of the dehydration-tolerant processes in plants. It would proceed at levels of water deficit sufficient to inhibit photosynthesis. The various measurements of dehydration tolerance are: maintenance of membrane integrity, plant growth, seedling survival, seedling growth, seed germination, osmotic adjustment, stem reserve metabolism and proline accumulation.

Breeding Methods

Following breeding methods are used for drought resistance:

1. **Introduction**: When drought resistance is available in exotic variety.
2. **Selection**: When drought resistance genotypes are available in land races or mixed population.
3. **Hybridization**: When resistant genes are available in germplasm or wild species. They can be transferred into cultivated varieties
4. **Mutation**: A desired gene is not present in the germplasm,but induced artificially.

The methods like introduction, pureline selection, mass selection, hybridization and mutation breeding in self-pollinated crops and mass selection, recurrent selection and synthetic varieties in cross pollinated crops are used.

Measurement of Drought Resistance

1. Leaf water retention

Leaves are excised from the genotypes and then dried. Those genotypes which exhibit slow drying are considered as drought tolerant.The tissue water potential is measured with the help of thermocouple psychrometer. The portable field psychrometer is widely used for measuring drought resistance in segregating population.

2. Rate of photosynthesis

The rate of photosynthesis during and after moisture stress is an important parameter of drought resistance. Portable non-destructive photosynthesis analyzers are used in the fields for large scale screening of germplasm as well as segregating population in standing crops. The genotypes which have high photosynthetic rate under moisture stress are considered as drought resistant.

3. Yield performance

Those genotypes which exhibit high yield performance under moisture stress conditions serve as an important reliable index of drought tolerance. The yield test should be conducted under both field as well as glass house conditions.

4. Roots length of seedling

The root length during seedling stage is also used as a measure of drought resistance. It is a rapid method of measuring drought resistance. Large number of plants can be screened for seedling root length in the green house during crop season.

Achievements

- **Rice**: Sahbhagi Dhan, Vandana, Anjali, Satyabhama, DRR Dhan 42 (IR64 Drt1), DRR Dhan 43, Birsa Vikas Dhan 203, Birsa Vikas Dhan 111, Rajendra Bhagwati.
- **Wheat**: PBW 527, HI 1531, HI 8627, HD 2888, HPW 349, PBW 644, WH 1080, HD 3043, PBW 396, K 9465, K 8962, MP 3288, HD 4672, NIAW 1415, HD 2987.
- **Maize**: Pusa Hybrid Makka 1, HM 4, Pusa Hybrid Makka 5, DHM 121, Buland.
- **Sorghum**: CSH 19 R, CSV 18, CSH 15R.

- **Pearl Millet**: HHB 67 improved, GHB 757, GHB 719, Dhanshakti, HHB Mandor Bajra Composite 2, HHB-226, RHB-177, Pusa Composite 443.
- **Barley**: RD 2660, K603.
- **Chickpea**: Vijay, Vikas, RSG 14, RSG 888, ICCV 10, Pusa 362.
- **Groundnut**: Ajaya, Girnar 1, TAG-24, Kadiri 6, ICGV 91114.
- **Soybean**: NRC 7, JS 95-60.
- **Sugarcane**: Co 98014 (Karan-1), Co 0239, Co 0118, Co 0238, Co 06927, Co 0403, Co 86032.
- **Cotton**: HD 324, CICR-1, Raj DH 7, Jawahar Tapti, Pratap Kapi, Suraj, Surabhi.
- **Mothbean**: RMO 60, 257, 435, 2251.

18

Disease Resistance

The disease , referred as the disorder of crop plants caused by fungi, mycoplasma, bacteria and viruses. Plant resistance is one of the most effective and ideal approach to combat diseases and insect-pest. Resistance is the capacity of plants to resist, withstand and overcome the attack of pathotypes.

Mechanism of Disease Resistance

Different mechanism of disease resistance are as under:

1. Disease escape

The ability of susceptible host plants to avoid attack of disease due to environmental factors like early maturing varieties, shift in the date of planting, change the field, balanced application of NPK etc. Example., early maturing varieties of groundnut and potato may escape from 'Tikka' and 'Late blight' diseases respectively.

2. Disease tolerance

The ability of the plants to tolerate the attack of the pathogen without loss or little loss in the yield. This endurance is brought about by the influence of external characters. Generally, tolerance is difficult to measure since it is confounded with partial resistance and disease escape. Disease tolerance is of three types:

1. **Disease resistance:** The ability of plants to withstand, oppose or overcome the attack of pathogens. There is infection and establishment, but development of pathogen in host tissue is restricted.
2. **Immunity:** When the host does not show the symptoms of disease, called as immune reaction. Pathogen may not reach the appropriate parts of the host due to immunity.
3. **Hypersensitivity:** Immediately after the infection several host cells surrounding the point of infection are so sensitive that they died. This leads to the death of the pathogen because the rust mycelium cannot grow through the dead cells. This super sensitivity (hypersensitivity) behaves as a resistant response for all practical purposes.

Genetic Basis of Disease Resistance

The term vertical and horizontal resistance was coined by Vanderplank. There are two main type of genetic basis of resistance:

1. Vertical resistance (VR)

VR is generally governed by major genes and is characterized by pathotype specificity. The immune or susceptible response in the case of vertical resistance depends on the presence of virulent pathotype. When virulent pathotype becomes frequent, epidemics are common. Thus an avirulent pathotype will produce an immune response i.e. r=0 or close to 0 but the virulent pathotype will lead to susceptible reaction i.e. r=1. It is also known as race specific, pathotype specific or simply specific resistance.

2. Horizontal resistance (HR)

This type of resistance is also called by various names like, race non-specific, pathotype-nonspecific and partial, or general resistance. Horizontal resistance is generally controlled by many genes i.e. polygenes with small effects and it is pathotype nonspecific. In this case, the reproduction rate is not zero but it is less than one. Poly genes, govern horizontal resistance. This type of resistance has low heritability, so identification of resistance is difficult. It provides protection from more number of races of pest.

Gene for Gene Hypothesis

The concept of gene for gene hypothesis between a host and its pathotype was first proposed by Flor in 1956 based on his work on linseed rust caused *by Malampsora lini*. The gene for gene hypothesis states that for each gene controlling resistance in the host, there is a corresponding gene controlling pathogenicity in the pathogen. The resistance of host is governed by dominant genes and virulence of pathogen by recessive genes. The genotype of host and pathogen determine the disease reaction. When genes in host and pathogen match for all the loci, then only the host will show susceptible reaction. If some gene loci remain unmatched, the host will show resistant. Now gene-for-gene relationship has been reported in several other crops like potato, sorghum, wheat etc. The gene for gene hypothesis is known as “Flor Hypothesis”.

Sources of Disease Resistance

There are different sources of resistance to diseases:

1. Cultivated variety

Disease resistance may occur in the cultivated variety. Resistant plants were also found from commercial varieties for ex., cabbage yellows in cabbage curly top resistance etc. These provide the basis for new resistance varieties.

2. Germplasm collection

Germplasm are reservoir of disease resistance almost in all the crops. Large number of germplasm should be collected from different sources and should be screened. In cotton several lines resistance to bacterial blight, and fusarium wilt were screened.

3. Related species

Resistance to a disease may be found in related wild species and transferred through interspecific hybridization. Example., Resistance to stem, leaf and stripe rusts of wheat.

4. Mutation

Spontaneous and induced mutations, both are used for development of disease resistance varieties. Example., Resistance to Victoria blight in oats was induced by irradiation with X-rays or thermal neutrons. Resistance to stripe rust in wheat.

Breeding Methods of Disease Resistance

Breeding methods for disease resistance are essentially same as those used for other agronomic traits. They are:

1. Introduction
2. Selection
3. Hybridization
4. Mutation Breeding.
5. Biotechnological methods.

Advantages of Disease Resistance

1. Reducing the losses caused by pathogens.
2. Reduces the high cost of disease control by chemical.
3. Helps to avoid the use of poisonous fungicides.
4. Only method available to some specific diseases likes viruses, wilt etc.
5. Avoid environmental pollutions.

Disadvantages of Disease Resistance

1. Linkage of resistant genes with genes of inferior quality.
2. Occurrence of physiological races of varying capacities.
3. Self sterility in host plants.

19

Variety Release Procedure

Crop breeders are continuously engaged in the improvement and development of crop varieties, in search of a genotype which will prove superior to the existence cultivars. A strain is duly released by appropriate duly empowered legal organization. Only when it is officially released, it is called variety, Before release, it is said to be strain or line. The release of new crop varieties consists of the following major steps:

1. Development of new strains

The new strains are developed by ICAR Crop Research Institutes and State Agricultural Universities for specific purposes. Various breeding methods are used for development of new strains. Some varieties are developed by selection without hybridization such as pure-line variety, mass selected variety and clonal variety. Development of some varieties involves hybridization and selection. This includes varieties that are developed by pedigree, bulk and backcross methods including multiline cultivars and population improvement methods.

2. Evaluation of performance

This trial is conducted at the research station, where new strains has breed. The object is to make sure that new strains are better in performance than the existence variety, so that it can be promoted to the trials where new strains has to be compete with other strains, developed by other scientists/researchers.

3. Multi-locational trials

The performance of newly developed strains is evaluated in AICCIP. ICAR Institutes, SAUs and private registered seed companies enter their improved strains/hybrids in the AICCIP of concerned crop for multi-location testing. The new strains are tested at multi-locations under the coordinated project for a minimum period of three years/ seasons to see their performance at different locations. The new variety is first tested for yield under the initial varietal trials (IET); the promising ones are promoted to the advanced varietal trials (AVT) in the second year. The strains that give good performance in AVT for two years are selected for further evaluation.

4. Agronomic trials

The strains performing well in AVT are included in the agronomic trials. Agronomic trials, determined its suitable dates of sowing, spacing, optimum number of irrigations, quantity of fertilizers, stage of fertilizers etc.

5. Adaptive trials

These trials are conducted on the state government's farm. Those entries which are identified in the workshop of the concerned coordinated project are included in adaptive trials.

6. Minikit trials

These trials are conducted at farmer's field in large areas.

7. Identification of superior strains

The strains which show good yield performance in AVT are identified as superior strains and are considered for release in the workshop meetings. The newer agro and plant protection techniques required to obtain potential yield of new strains are also worked out by that time. The workshop after considering the new promising varieties, recommend them to replace existing varieties.

8. Release and notification

The purpose of release of cultivars is to introduce the newly evolved varieties to the public for general cultivation in the region in which it is suitable. It enables the farmers to choose cultivars for cultivation in a region. The proposal for release of new varieties is put up in a prescribed proforma to variety release committee. There are two types of variety release committees, viz. State Variety Release Committee (SVRC) and Central Variety Release Committee (CVRC). In case of state variety release committee, Director of Agriculture of the State Department of Agriculture is the chairman (SVRC). In central variety release committee, Deputy Director General (Crop Science) of ICAR is the Chairman. The release proposal of varieties recommended for All India release is put up before CVRC, while for those recommended for release in a particular state is placed before the SVRC of respective state. These committees consist of scientists, and representatives of seed producing organizations (NSC, SSC and SSCA) and other related government agencies. After release, the variety is notified. Seed production can be taken up only after notification of new varieties. The notification is done by the government of India.

9. Notification

Notification of cultivars after official release (at State as well as Central levels), the cultivars are notified under the Seeds Act, so that the quality of seeds can be regulated. The main purpose of notification is to bring the seeds of a particular crop/variety under the purview of Seed Law Enforcement, mainly to empower the Seed Inspectors to verify the quality of its seeds by sampling and analysis. The notification is made by the Central Government on the recommendation of the Central Seed Committee. The proposals for notification of a State-important variety are forwarded in the prescribed format by the State government after its release in that State to the Central Seed Committee for consideration.

Procedure for release of a variety

20

Classes of Seed

The prime objective of any plant breeder is to develop high yielding varieties and make the quantity of quality seed of those varieties to farmers for commercial cultivation to realize the benefits of superior varieties. This necessitates, the production of seed of improved varieties in large scale in order to make them available to farmers without losing their genetic purity. Seed production is a continuous process. The seed of improved varieties are produced in several stages, each class producing a special class of seed. The various classes of improved seeds are recognized in order to facilitate seed production and to ensure continuous supply of quality seed. A seed production programme for the production of quality seed has different classes of seeds viz; nucleus seed, breeder seed, foundation seed, registered seed and certified seed. The Association of Official Seed Certifying Agencies (AOSCA) has defined these seed classes as follows:

1. Nucleus Seed

Basic or nucleus seed is the original seed of a released and notified variety. Nucleus seed is multiplied by single plant progeny system, using the concept of maintenance breeding, where purity of the original parental stock is maintained. It is available in very limited quantities and taken from the individual plant of a variety for purification. It is directly under the controlled of concerned crop breeder. This seed has hundred percent genetic and physical purity. Breeder seed is multiplied out of this material and about 500-1000 plants are kept for continued multiplication of nucleus seed.There is no need of any type of tag on nucleus seed but concerned breeder issue a certificate for purity.

2. Breeder Seed

Breeder seed is the progeny of nucleus seed which is directly controlled of the originating or the sponsoring breeder or institution. Breeder seed is maintained by growing the single plant progeny and monitoring the production programme contineouly. If any suspection, removable of whole plant progeny rows. Breeder seed has cent percent genetic and physical purity. This type of seed does not require any type of certification but inspected by monitoring team for maintain the quality of seed. Breeder seed tag is golden yellow. It is not used for general distribution. This is the source of foundation seed production.

3. Foundation Seed

This is the progeny of breeder seed. The seed stock handled to maintain specific identity and genetic purity, which may be produced at research farms of agricultural university, KVK, government farms under strict supervision of breeder and NSC. This class of seed is certified by SSC Agency. White tag is attached with the seed bag. This seed is the source of all other certified seed classes either directly or through registered seed.

4. Registered Seed

The progeny of the foundation seed so handled as to maintain its genetic identity and purity and approved and certified by a certifying agency. This type of seed is not used in India.

5. Certified Seed

Certified seed is the progeny of the foundation or certified seed. Its production is so handled to maintain satisfactory genetical identity and physical purity according to standards specified for the crop being certified. This is the commercial seed which is available to the farmers and its genetic purity should be 99 per cent. Certified seed is produced by SSC, NSC, SFCI, Agriculture university, Government farms, seed companies and progressive farmers. Seed is certified by SSC Agency. Blue tag is used with seed bag

6. Truthful Labelled Seed

The seed of notified variety which is not up to the standard of certification but full fill the seed standard up to good label. This category of seed does not required seed certification.

Steps of production of different classes of Seed

Nucleus seed stage I ___________ Nucleus seed stage II _____________
Breeder seed stage I _________ Breeder seed stage II _____________
Foundation seed stage I _________ Foundation seed stage II ____________
Certified seed stage I _______________ Certified seed stage II ____________

21

Crop Ideotypes

The term ideotype was proposed by Donald in 1968. Ideotype is a biological model which is expected to perform in a predicted manner in a defined environment. A crop ideotype is a plant model that is expected to perform better in a particular environment. The concept of ideotype changes from crop to crop, climate to climate and the concepts and policies of the community concerned. Ideotype breeding is the development of a plant type, expected more quantity of grains, with ideal character. Plant characters are selected from an agronomic point of view.

Characteristics of Ideotype

1. Model plant for a specific environment.
2. Ideotype are the most efficient in utilizing its environmental resources.
3. Ideotype is a moving goal.
4. The design of crop ideotypes is likely to involve concurrent modification of the environment.
5. Ideotype must include such morphological and physiological characteristics that result in a high harvesting index.
6. A crop ideotype must be grown, as far as possible, in weed – free situation in view of it being a weak competitor.
7. Model plants are expected to give more yield than the old variety.

Features of Crop Ideotype

Crop ideotype consists of several morphological and physiological traits which enhance the yield potential through genetic improvement of individual characters. Morphological and physiological features crop ideotype vary from crop to crop or sometimes differ within the crop also depending upon weather. The ideotype are required for irrigated or rainfed situation. The important ideotype for some crops are given below:

1. **Wheat:** The ideotype on wheat was first decribed by Donald in 1968. Ideotype of wheat with following main features. i) A short strong stem, ii) Erect leaves, iii) Few small leaves, iv) Larger ear, v) Presence of owns, vi) A single culm, vii) Effective weed control, viii) Suitable planting density, ix) Short plant height, x) Free from diseases.

2. **Rice:** The concept of plant type was introduced in rice breeding by Matsushima in 1957 and latter by Jennings in 1964. Chandler in 1969 suggested that the rice an ideal or model plant type consists of i) Semi dwarf stature, ii) 4-5 productive tillers, iii) Dark, green,short, erect, thick and highly angled leaves, iv) Short culm length, v) Vigorous root system, vi) No lodging, vii) Large panicle.

3. **Maize:** Mock and Pearce in 1975 proposed ideal plant type of maize. In maize, higher yields can be obtained from the plants consisting of i) Low tillers, ii) Large cobs, iii) Angled leaves for good light interception, iv) Short interval between pollen shed and silk emergence, v) Small tassel size, vii) Long grain filling period, viii) Planting of such type at closer spacing resulted in higher yields.

4. **Barley:** Rasmusson (1987) reviewed the work on ideotype breeding and also suggested ideal plant type of six rowed barley. i) Short stature, ii) Long awns, iii) High harvest index, iv) High biomass, v) Resistant to diseases, vi) More spike length.

5. **Cotton:** Singh et al. (1974) proposed ideal plant type of upland cotton growing belt. i) Short stature (90-120 cm), ii) Compact and sympodial plant habit, iii) Short duration (150-165 days), v) Responsive to high fertilizer dose, vi) High degree of inter plant competitive ability,vii) Resistance to insect pests and diseases, viii) More physiological efficiency, ix) Boll size between 3.5-4g.

6. **Sorghum:** Swaminathan (1972) has proposed several desirable attributes of crop ideotype with special reference to multiple cropping in the tropics and sub tropics. i) Higher population, ii) High productivity per day, iii) More photosynthetic ability, iv) Low photo respiration, v) Low photo and thermo sensitivity, vi) High response to nutrients, vii) High productivity per unit of water, viii) Multiple resistances to insect and diseases.

7. **Chickpea:** Ideal plant type for rainfed conditions: i) Early vigour, ii) 50-60 cm plant height with 9-10 secondary branches, iii) Tall, erect or semi-erect plant, iv) More number of pods per plant, v) Podding from 10th node, vi) Early in maturity, vii) Well developed root system. For irrigated conditions: i) High input responsiveness, ii) Tall (75-90 cm) and erect habit with broom shaped branching behavior, iii) Resistance to insect-pest and diseases.

8. **Pigeonpea:** i) Synchronous flowering, ii) delayed senescence, iii) determinacy, iii) Long fruiting branches and short inter nodes, iv) Lodging resistance, v) Pod bearing from 20 cm above the ground, vi) Semi-dwarf plant type (1.5 – 1.8 m) for mechanized plant protection, vii) Open canopy with deteminancy, viii) Non-cluster pod bearing, ix) Long fruiting branches for high yield, x) Plant type for intercropping.

9. **Moong bean:** For kharif conditions: i) Optimum duration, ii) Balanced vegetative growth, iii) Clear distinction between vegetative and reproductive phase, iv) Tall plants (80-100 cm) with more branches, v) Synchronous maturity, vi) More no. of clusters/plant and pods, vii) More number of seeds/ pod, viii) Shattering and pre-harvest sprouting tolerance. For summer conditions: i) Shorter duration (50-60 days), ii) Medium plant height (60-80 cm), iii) Determinate growth habit and synchronous maturity, iv) More number of pods at top of plant, v) Non-shattering type, vi) Longer pods, vii) Tolerance to terminal heat stress.

10. **Urd bean:** i) Optimum maturity (65-75 days), ii) Determinate growth habit, iii) Upright plant growth habit, iv) More plant height, v) More number of clusters/plant and pods/cluster, vi) Synchronous maturity.

11. **Lentil:** i) Optimum duration, ii) Compact type with strong collar region and stiff stem, iii) Tall (50-60 cm) and erect plant, iv) Lodging resistance, v) bearing from 8th node, vi) Prolific root system, vii) Pods borne well above the soil surface, viii) Reduced pod dehiscence ,ix)Large seeds, x) Response to inputs, xi) Tolerance to heat and cold.

12. **Mothbean:** i) Early in maturity (62-65 days, ii) Profuse branching, iii) Well developed root system, iv) Tolerance to drought, v) More number of seeds per pod.

13. **Brassica spp:** i) Height of 1-1.25m, ii) 5-6 primary branches at an angle of 40-45° iii) Main stem should bear only 40 siliqua, iv) Each siliqua should have 20 seeds, v) Seed weight should be 5mg, vi) Deep root system, vii) Thicker leaves with high photosynthetic rate and nitrate reductase activity. viii) Shattering resistance, ix) Pest and diseases resistant.

Merit of Ideotypes

1. It exploits both morphological and physiological variation.
2. Various morphological and physiological traits are specified and each character or trait contributes towards high yield.
3. Involves research experts from different discipline. Each researcher contributes for the development of model plants on which he/she is working.

4. Ideotype breeding is an effective method of breaking yield barriers through the use of various improved characters that contributing towards higher yield.
5. This approach provides solution to various problems related to disease, insect-pest, lodging resistance, maturity duration, yield and quality by transferring desirable genes from other sources into a single genotype.
6. It is efficient method of developing varieties for specific or environment.

Demerits of Ideotypes

1. It is a difficult task to combine various features from different sources in to a single genotype.
2. Ideotype breeding is a time consuming method of cultivar development.
3. Through ideotype breeding, traditional or conventional breeding can not be substituted. It is only a supplement.
4. New ideotype have to evolve to meet the changing and increasing demands of economic products.

22

Plant Tissue Culture

Tissue culture is a technique used to culture cells, tissues, organs or whole plant under controlled environmental conditions that *in vitro* condition is known as plant tissue culture. Plant tissue culture exploits the totipotency nature of plant cells, ability to develop in to complete plantlet.

Landmarks of Tissue Culture

- **Gottlieb Haberlandt 1902:** For the first time attempted to culture single cells. He is also known as the father of plant tissue culture.
- **Haberlandt 1902:** Gave the concept of *in vitro* cell culture.
- **Hannig- 1904:** Cultured embryos from several cruciferous species.
- **Went 1926:** Discovered first plant growth hormone that Indole acetic acid.
- **Overbeek 1941:** First time add coconut milk for cell division in Datura.
- **Skoog and Miller 1955:** Discovered kinetin as cell division hormone.
- **Skoog and Miller 1957:** Gave the concept of hormonal control (auxin: cytokinin) of organ formation.
- **Reinert and Steward 1959:** Regenerated embryos from callus clumps and cell suspension of carrot *(Daucus carota).*
- **Cocking 1960:** Isolated protoplast by enzymatic degradation of cell wall.
- **Kanta and Maheshwari 1960:** Developed test tube fertilization technique.
- **Murashige and Skoog 1962:** Developed MS medium with higher salt concentration.
- **Guha and Maheshwari 1964:** Produced first haploid plants from pollen grains of Datura (Anther culture).
- **Power *et al.*, 1970:** Successfully achieved protoplast fusion.
- **Carlson 1972:** Developed first interspecific hybrid of *Nicotiana tabacum* by protoplast fusion.
- **Chilton *et al.*, 1977:** Successfully integrated Ti plasmid DNA from *Agrobacterium tumefaciens* in plants.

- **Melchers *et al.*, 1978:** Carried out somatic hybridization of tomato and potato resulting in pomato.
- **Larkin and Scowcroft 1981:** Introduced the term somaclonal variation.
- **Horsh *et al.*, 1984:** Developed transgenic tobacco by transformation with Agrobacterium.
- **Klien *et al.*, 1987:** Developed biolistic gene transfer method for plant transformation.

Advantages of Tissue Culture

1. Production of improved crop varieties.
2. Production of disease-free plants.
3. Genetic transformation.
4. Production of secondary metabolites.
5. Production of varieties tolerant to salinity, drought and heat stresses.

Plant Tissue Culture: Laboratory and Equipments

Plant tissue culture laboratory has several functional areas. The different area of plant tissue culture laboratory includes:

1. Media preparation room
2. Washing/Sterilization area
3. Inoculation room/Sterile transfer area
4. Culture room/Incubation room/Growth room
5. General storage area
6. Research area
7. Planting area

The range of equipments needed for laboratory depends on the research work. Major equipments, required for laboratory are:

1. Laminar air flow
2. Autoclave
3. Hot air oven
4. pH meter
5. Lux meter
6. Hygrometer
7. Distillation unit
8. Shaker

9. Microscope
10. Heater
11. Magnetic stirrer

Plant Tissue Culture Media

Plant tissue culture medium contains all the nutrients essential for the normal growth and development of plants. Medium is generally composed of macronutrients, micronutrients, vitamins, other organic components, plant growth regulators, carbon source and some gelling agents in case of solid medium. Murashige and Skoog medium (MS medium) is most commonly used for the vegetative propagation of many plant species *in vitro*. The pH of the media is also important that affects both the growth of plants and activity of plant growth regulators. It is adjusted between 5.4 - 5.8. Both the solid and liquid medium can be used for culturing. The most commonly used medium are MS medium, white medium, Gamborg B5 medium, Nitch and Nitch medium, SH (Shank and Hilderbrant), WPM (Llyod and Mccown), N6 etc. Plant growth regulators (PGR's) play an essential role in determining the development pathway of plant cells and tissues in culture medium. The auxins, cytokinins and gibberellins are most commonly used plant growth regulators. Auxins and cytokinins are most widely used plant growth regulators in plant tissue culture and their amount determined the type of culture established or regenerated. The high concentration of auxins generally favours root formation, whereas the high concentration of cytokinins promotes shoot regeneration. A balance of both auxin and cytokinin leads to the development of mass of undifferentiated cells known as callus

The Principle Component of Plant Tissue Culture Media

1. Inorganic nutrients (macro and micro)

The mineral nutrient required in concentrations grater than 0.5 mmols/l are referred as macronutrients (N, P, K, Ca, Mg and S) and those required in concentrations less than 0.5mmols/l are referred as micronutrients (Fe, B, Cl, Co, Cu, Mn, Zn, KI).

2. Organic supplements

Vitamins like, thiamine, nicotinic acid and pyridoxine.

3. Growth regulators

i) **Auxins**: Induces cell division, cell elongation, swelling of tissues, formation of callus formation of adventitious roots. Inhibits adventitious and axillary shoot formation. Example., 2,4-D, NAA, IAA, IBA.

ii) **Cytokinins:** Shoot induction, cell division. Example., BAP, Kinetin, zeatin, 2iP.

iii) **Gibberellins:** Plant regeneration, elongation of internodes. Example., GA3, Abscisic acid: induction of embryogenesis.

iv) **Abscisic acid:** Inhibit growth habit, useful in embryo culture, heat stable but light sensitive.

4. Carbon source

Glucose that enhances proliferation of cells and regeneration of green shoots.

5. Gelling agents

The most commonly used gelling agent is agar-agar obtained from seaweed *Gelidium amansii.*

Sterilization Techniques

Plant tissue culture media, which contain high concentrations of sugars, support the growth of many microorganisms like bacteria and fungi. The initial explant is the major source of infection, but re-infection is possible at any stage of culture process. The maintenance of aseptic or sterile conditions are therefore, absolutely essential to maintain a completely aseptic environment for successful tissue culture procedures. Generally, four categories of sterilization are recognized:

1. Sterilization of Culture Vessels and Instruments

i) **Dry heat sterilization:** Glass, culture vessels, metal instruments and aluminium foil can be sterilized by exposure to hot dry air at 160-180°C for 2-4 h in a hot air oven.

ii) **Autoclaving:** Cotton plugs, gauze, plastic caps, or pipettes are autoclaved at 121°C for 15-20 minutes.

iii) **Flame sterilization:** Normally metal instruments such as forceps, scalpel, needles, and spatulas are sterilizing by dipping in 95% alcohol followed by flaming and cooling.

2. Sterilization of Nutrient Media

i) **Autoclaving:** Culture media in glass containers sealed with cotton plugs, aluminium foil at 15 psi for 15-40 minutes.

ii) **Filter sterilization:** Vitamins, amino acids, hormones, and carbohydrates are thermolabile and may decompose during autoclaving. Hence, these are sterilized by filter sterilization.

3. Sterilization of explants

Surface of plant parts carry a wide range of microbial contaminants. To avoid the source of infection, the tissue must be thoroughly surface sterilized by calcium hypochloride or sodium hypochloride or mercuric chloride or hydrogen peroxide etc.

4. Sterilization of Culture Room and Transfer area

Laminar flow hood, which is provided with UV sterilizing tubes. The transfer area is cleaned once or twice a month with a commercial brand of a disinfectant. However, larger transfer rooms are best sterilized by exposure to UV light. All the sterile transfer is to be done in laminar flow.

Types of Tissue Culture

Different types of plant tissue cultures techniques are summarized as follows:

1. Callus culture

Callus is an undifferentiated and unorganized mass of parenchyma tissue. In living plants, callus are undifferentiated mass cells which cover a plant wound. The culture medium is supplemented with a defined amount of plant growth regulators such as auxins, cytokinins and gibberellins to initiate callus formation or formation of somatic embryos. During this process cell differentiation and specialization is reversed and the explants gives rise to new tissue. The callus which is formed on an original explant is called primary callus.

2. Suspension culture

This is a type of culture in which single cells/protoplast or small cell aggregates multiply in agitated liquid medium. The cells remain suspended in a defined aseptic media, controlled physiological condition and regulated environmental condition. It is also referred to as cell culture or cell suspension culture. Suspension culture grows much faster than the callus cultures. Several methods of suspension cultures have been developed. The two main types are: Batch culture-in which the cells are nurtured in a fixed volume of medium until growth ceases and in continuous culture -in which cell growth is maintained by continuous replenishment of nutriment medium.

3. Meristem culture

Meristematic tissue is present at several locations in the plants. In roots and shoots, meristematic tissue is categorized into three groups: Apical meristem: Present at the root and shoots tips of the plant. The divisions in this area facilitate growth and development of root and shoot. Intercalary meristem: Located at the intercalary position and helps to increase the height of the internodes. Lateral meristem: Located at the lateral side of stems and roots, involved in increasing the thickness of the plant.

Procedure of Meristem Culture

1. **Explant:** (i) Shoot tip of 100-500 μm with 1-3 leaf primordia. (ii) Shoot tip up to 1mm for virus elimination. (iii) 5-10 mm explants for clonal propagation.
2. **Culture medium:** Placed the exercised meristem on MS medium is satisfactory but in some species with lower salt concentration is desired. Grow the plantlet from meristem to the six node stage.
3. **Growth regulators:** GR free medium is used, if heavily contaminated then bavistin may be added.
4. **Shoot multiplication:** After 2-3 weeks, the cultures are transferred to shoot multiplication medium containing cytokinin to promote axillary branching.
5. **Environment during culture:** During culture initiation and shoot multiplication phases, they are kept at 25°C are illuminated with 1000 lux white light.
6. **Rooting of shoots:** It has low salts and reduced sugar levels
7. **Transfer of plantlets to soil:** Plants are kept in high humidity and low light

Applications of Meristem Culture

1. Production of virus free plants.
2. Mass production of desirable genotypes.
3. Facilitation of exchange between locations (production of clean material).
4. Cryopreservation (cold storage) or in vitro conservation of germplasm.
5. Meristem culture is useful in breeding programmes where hybrid plants produce non-viable or abortive seeds and are unable to mature. The meristem and shoot tips have also some application in Agrobacterium mediated genetic.

4. Anther culture

Anther culture is a technique by which immature anther is made to divide and grow into tissue (either callus or somatic embryo), primarily to produce haploids plants with (n) number of chromosome/ haploid set of chromosome.

5. Pollen culture (microspore culture)

Pollen culture is a technique in which haploid plants are obtained from isolated pollen grains. The pollen grains are dusted on a suitable culture medium and allowed to grow into callus/somatic embryo. Plants can be regenerated through shoot and root induction in haploid callus and through culture of haploid embryos. It is possible to produce homozygous, doubled haploid, pure breeding lines through chromosome doubling treatments applied to haploid cells.

6. Ovary/ovule culture

Ovules are aseptically isolated from the ovary and are grown aseptically on chemically defined nutrient medium under controlled conditions. Ovary culture is a technique of culture of ovaries isolated either from pollinated or un-pollinated flowers. When the tissue of pollinated and fertilized ovary is taken, it will produce diploid plants and when unfertilized ovary tissues are taken after it has gone meiosis, it will result in the formation of haploid plants. There is a scarcity of haploid cells within an ovule as compared to the anther, so anther or microspore culture is generally used to develop haploid plants.

7. Embryo culture

Embryo culture is a type of plant tissue culture that is used to grow embryos from seeds and ovules in a nutrient medium. In embryo culture, the plant develops directly from the embryo or indirectly through the formation of callus and then subsequent formation of shoots and roots. The technique has been developed to break seed dormancy, test the vitality of seeds, production of rare species and haploid plants. It is an effective technique that is employed to shorten the breeding cycle of plants by growing excised embryos and results in the reduction of long dormancy period of seeds. Intra-varietal hybrids of an economically important energy plant.

8. Protoplast culture

Protoplast fusion provides an efficient mean of gene transfer with desired trait from one species to another and has an increasing impact on crop improvement. Somatic hybridization is an important tool of plant breeding and crop improvement by the production of interspecific and intergeneric hybrids. The culture of protoplast in an aseptic liquid media is known as protoplast culture. Protoplast is a cell without a cell wall. It is a living part of a plant cell consisting of cytoplasm and nucleus. The basic principle of protoplast culture is the aseptic isolation of large number of living protoplasts and cultures them on a suitable nutrient medium for their requisite growth and development. Protoplast can be isolated from variety of plant tissues but best and maximum number of protoplast can be isolated from mesophyll tissue of leaves. Single cell suspension culture can be considered as protoplast culture. Protoplast maybe fused together to form hybrids and cybrids. Protoplast culture is used to introduce and modify genetic information inside any cell. It is used to produce transgenic plants.

9. Somatic embryogenesis

Embry that can originate from different types of somatic cells also. These type of embryos are known as somatic embryos and the process of formation of somatic embryo is known as somatic embryogenesis. Somatic embryos are a single cell or a group of cells capable of regeneration into a complete plant. Somatic embryos can be produced directly or indirectly. In direct method embryos are produced from cell or group of cells without forming callus. In indirect method firstly callus are produced from somatic tissues and then embryos are produced from callus tissues or from cell suspension produced from that callus. These embryos are bipolar in nature i.e. they have a radical and a plumule. They are used in synthetic seed formation, genetic transformation, mass multiplication of elite germplasm and can be used as materials of embryogenic protoplast.

Micropropagation

Vegetative propagation *in vitro* is called as micropropagation. A general description of these consists of the following stages:

1. Selection of mother plant

Careful attention should be given to the selection of the parent plants. They must be variety or species, and free from disease. Treat the selected plant with chemical sterilizing agents like, sodium hypochloride, calcium hypochloride, etc.

2. Establishing an aseptic culture

Success of this stage primarily requires that explants should be transferred to the cultural environment, free from microbial contaminants; and this should be followed by some kind of growth Success of this stage depends on adequate number of explants had survived without contamination, and kept growing.

3. Production of suitable propagules

Production of new plant outgrowths or propagules which, when separated from the culture are capable of giving rise to complete plants.

4. Preparation for growth in the natural environment

Plantlets are then transplanted into an adequate rooting medium (such as a peat: sand compost) and kept for several days in high humidity and reduced light intensity. A fog of water vapour is very effective for maintaining humidity. Alternatively, intermittent water misting may be applied automatically, or the plants may be placed inside a clear plastic enclosure and misted by hand.

Advantages of Micropropagation

1. Requirement of explant is minimal.
2. Faster rate of propagation and many more plants can be produced in a short period of time.
3. The new product is free from viral diseases.
4. Possible to produce clones of rare species/genotypes, and also plants that is otherwise slow and difficult to propagate vegetatively.
5. Plants may acquire a new characteristic through micropropagation.
6. Production can be continued all the year.
7. Vegetatively-reproduced material can often be stored over long periods.
8. Plant material needs little attention between sub cultures and no labour requirement for watering, weeding, spraying etc.

Disadvantages of Micropropagation

i) The cost of propagules is usually relatively high.

ii) Required sophisticated equipment and technical staff.

iii) Required well equipped laboratory.

iv) Techniques required for maintaining stringent aseptic environment.

v) Large number of chemicals and glasswares are required.

vi) Required green house.

vii) Plantlets obtained are initially small and sometimes have undesirable characteristics caused by somaclonal variation.

Somaclonal Variation

Somaclonal variation is defined as genetic variation observed among progeny of plants regenerated from somatic cells cultured *in vitro*. Somaclonal variation is the genetic variations produced in plants in response to plant tissue culture or any form of cell culture and the term somaclonal was coined by Larkin and Scowcroft (1981). Somaclonal variation can also be a useful tool for crop improvement through selection and characterization of novel variations. The successful example of such variations is a Fiji disease resistant variety 'ono' developed through somaclonal variation in sugarcane from variety Pindar; and a bold seeded and high yielding variety Pusa Jai Kisan developed in Indian mustard from variety Varuna. Another example of somaclonal scarlet is a sweet potato which has darker and more stable skin colour

Advantages of Somaclonal Variations

1. Used in crop improvement.
2. Used to generate genetic variations.
3. Used to increased and improved production of secondary metabolites.
4. Selection of plants resistant to toxins, herbicides, high salt concentration and mineral toxicity.
5. Suitable for breeding of tree species.

Disadvantages of Somaclonal Variations

1. Require clonal uniformity.
2. Sometime leads to undesirable results.
3. Selected variants are random and genetically unstable.
4. Required prolonged field trials.
5. Not suitable for traits like yield, quality etc.
6. May develop variants with pleiotropic effects which are not true.

23

Genetic Engineering

Advances in molecular biology have sharpened the tools of the breeders, and brighten the prospects of confidence to serve the humanity. The application of biotechnology to field crop has already led to the field testing of genetically modified crop plants. Genetically engineered Rice, Maize, Soybean, Cotton, Oilseeds like Rape seed, Sugar Beet and Alfalfa cultivars are expected to be commercialized in the late of 20th century. Genes from varied organisms may be expected to boost the performance of crops especially with regard to their resistance to biotic and abiotic stresses

Genetic engineering involves isolation of DNA fragments within a cell recombining them outside. The basic technique is quite simple. Two DNA molecules are isolated and cut into fragments by one or more specialized enzymes, the fragments joined together in a desired combination, then restored to a cell from replication and reproduction. This method also referred to as recombinant DNA technology. Genetic engineering developed in the mid 1970 when it becomes possible to cut DNA and to transfer particular pieces of DNA containing specific bits of information, from one type of organism into other.

Recombinant DNA Technology

DNA cloning involves separating a specific gene or DNA segment from a larger chromosome, attaching it to a small carrier DNA. The resultant hybrid DNA is called recombinant DNA, which is transferred to a proper host (bacteria, virus or yeast) and replicated to make multiple copy of the selected gene. Recombinant DNA technology was developed by Boyer and Cohen in 1973.

Essential Components for Gene Cloning

1. Enzymes for cutting and joining the DNA fragments

Restriction endonucleases cut at sites within the foreign DNA instead of chewing from the end. By cutting DNA at specific sites they function as finely honed molecular knives.

Type II restriction enzyme nomenclature

- *Eco*RI - *Escherichia coli* strain R, 1st enzyme
- *Bam*HI - *Bacillus amyloliquefaciens* strain H, 1st enzyme
- *Dpn*I - *Diplococcus pneumoniae*, 1st enzyme
- *Hind*II - *Haemophilusinfluenzae*, strain D, 2nd enzyme
- *Bgl*II - *Bacillus globigii*, 2nd enzyme
- *Pst*I - *Providencia stuartii*164, 1st enzyme
- *Sau*3AI - *Staphylococcus aureus* strain 3A, 1st enzyme
- *Kpn*I - *Klebsiella pneumoniae*, 1st enzyme

2. Vectors or Cloning Vehicles

A vector is a DNA molecule that has the ability to replicate in an appropriate host cell, and into which the DNA insert is integrated for cloning. The vector is a vehicle (or) carrier which is used for cloning foreign DNA in bacteria.

Properties of a good vector

1. Able to replicate.
2. Easy to isolate and purify.
3. Easily introduced into the host cells.
4. The vector should have suitable marker genes that allow easy detection or/ and selection of the transformed the host cell.
5. The cells transform with recombination DNA should be identifiable.
6. A vector have unique target sites for as many restriction enzymes as possible into which the DNA insert can be integrated.
7. When expression of the DNA insert is desired, the vector should contain suitable regulatory elements like promoter, operator and ribosome binding sites.

Types of Vector

- Plasmid vector
- Bacteriophage vector
- lambda-phage vector
- Cosmid vector
- Shuttle vector
- Yeast vector
- Plant vector

3. DNA Fragments

The isolation of DNA fragment to be cloned is a critical step in gene cloning. DNA insert can be obtained from:

1. Restriction endonuclease

Restriction enzymes are endonucleases produced by bacteria that typically recognize specific 4 to 8bp sequences, called restriction sites, and then cleave both DNA strands at this site. Restriction sites commonly are short *palindromic* sequences; that is, the restriction-site sequence is the same on each DNA strand when read in the 5'→ 3' direction.

2. Amplification through PCR

The Polymerase Chain Reaction (PCR) has the following steps:

i) Template DNA

ii) Two oligonucleotide primers (Forward & Reverse)

iii) Four deoxynucleoside triphosphates (TTP, dCTP, dATP and dGTP)

iv) Heat stable DNA polymerase (T*aq*)

v) Denaturation – When the double-stranded template DNA is heated (94° to 96 °C) to separate it into two single strands

vi) Annealing – When the temperature is lowered (50° to 56°C) to enable the DNA primers to attach to the template DNA

vii) Extending – When the temperature is raised (72° to 75°C) and the new strand of DNA is made by the Taq polymerase enzyme.

3. DNA libraries

A DNA library is a collection of clones of DNA designed so that there is a high probability of finding any particular piece of the source DNA in the collection. It is of two types:

i) **Genomic library:** The genomic library contains DNA fragments representing entire genome of an organism.

ii) **cDNA library:** The cDNA library contains only complementary DNA molecules synthesized from mRNA molecules in a cell.

4. Selection of a clone of transformed cells that has acquired the recombinant chimeric DNA molecule

Steps of Gene Cloning

1. Production of DNA fragments to be cloned. Cutting the DNA to be cloned from the specific restriction endonuclease.
2. Insertion of DNA fragments in to a vector.
3. Incubating the vector and subject DNA togather to anneal and then joining them use DNA ligase. The resultant DNA is called recombinant DNA.
4. Transferring the reconbinant DNA to an appropriate host such as bacteria, virus or yeast which will provide necessary biomachinary for DNA replication.
5. Identifying the host cells that contain the recombinant DNA.

Application

1. Gene cloning plays important role in the production of antibiotics, hormones, vaccines and interferon in the field in medicine.
2. Role of gene cloning in identification of genes responsible for human diseases.
3. Production of recombinant insulin.
4. Distant hybridization.
5. Development of transgenic plants.
6. Development of nodules in cereals.
7. Development of c4 plants.

Blotting Techniques

1. **Southern blotting:** Southern blotting techniques is the first nucleic acid blotting procedure developed in 1975 by Southern. Southern blotting is the techniques for the specific identification of DNA molecules.
2. **Northern blotting:** Northern blotting is the techniques for the specific identification of RNA molecules.
3. **Western blotting:** Western blotting involves the identification of proteins.

Transgenic Plants

Plants in which foreign DNA has been artificially incorporated into its genome. Mostly genetic modified traits are used as herbicide resistance traits followed by quality improvement and insecticide resistance traits.

Methods of Gene Transfer

1. Agrobacterium

Vector-mediated gene transfer is carried out by Agrobacterium-mediated transformation. *A. tumeifaciens* causes crown gal disease, *A. rhizogenes* causes hairy root disease. *Agrobacterium tumefaciens* is a soil-borne, gram-negative bacterium. It is rod shaped and motile, and belongs to the bacterial family of Rhizobiaceae. *A. tumefaciens* is a phytopathogen, and is treated as the nature's most effective plant genetic engineer. As *A. tumefaciens* infects wounded or damaged plant tissues, in induces the formation of a plant tumor called crown gall. The entry of the bacterium into the plant tissues is facilitated by the release of certain phenolic compounds (acetosyringone, hydroxyacetosyringone) by the wounded sites. Crown gall formation occurs when the bacterium releases its Ti plasmid (tumor-inducing plasmid) into the plant cell cytoplasm. A fragment (segment) of Ti plasmid, referred to as T-DNA, is actually transferred from the bacterium into the host where it gets integrated into the plant cell chromosome (i.e. host genome). Thus, crown gall disease is a naturally evolved genetic engineering process

Among the various vectors used in plant transformation, the Ti plasmid of *Agrobacterium tumefaciens* has been widely used. This bacteria is known as "natural genetic engineer" of plants because these bacteria have natural ability to transfer T-DNA of their plasmids into plant genome upon infection of cells at the wound site and cause an unorganized growth of a cell mass known as crown gall. Ti plasmids are used as gene vectors for delivering useful foreign genes into target plant cells and tissues. The foreign gene is cloned in the T-DNA region of Ti-plasmid in place of unwanted sequences. To transform plants, leaf discs (in case of dicots) or embryogenic callus (in case of monocots) are collected and infected with *Agrobacterium* carrying recombinant disarmed Ti-plasmid vector. The infected tissue is then cultured on shoot regeneration medium for 2-3 days during which time the transfer of T-DNA along with foreign genes takes place. After this, the transformed tissues (leaf discs/calli) are transferred onto selection cum plant regeneration medium supplemented with usually lethal concentration of an antibiotic to selectively eliminate non-transformed tissues. After 3-5 weeks, the regenerated shoots (from leaf discs) are transferred to root-inducing medium, and after another 3-4 weeks, complete plants are transferred to soil following the hardening of regenerated plants. The molecular techniques like PCR and southern hybridization are used to detect the presence of foreign genes in the transgenic plants.

2. Direct Gene Transfer

In these methods, the foreign gene of interest is delivered into the host plant cell without the help of a vector. The methods used for direct gene transfer in plants are:

1. **Chemical mediated gene transfer:** Chemicals like polyethylene glycol (PEG) and dextran sulphate induce DNA uptake into plant protoplasts. Calcium phosphate is also used to transfer DNA into cultured cells.
2. **Microinjection:** DNA is directly injected into plant protoplasts or cells (specifically into the nucleus or cytoplasm) using fine tipped (0.5 - 1.0 μm) glass needle or micropipette. This method of gene transfer is used to introduce DNA into large cells such as oocytes, eggs, and the cells of early embryo.
3. **Elecrtroporation:** Involves a pulse of high voltage applied to protoplasts/ cells/ tissues to make transient (temporary) pores in the plasma membrane which facilitates the uptake of foreign DNA. The cells are placed in a solution containing DNA and subjected to electrical shocks to cause holes in the membranes. The foreign DNA fragments enter through the holes into the cytoplasm and then to nucleus.
4. **Particle gun/Particle bombardment:** Particle bombardment is a technique used to introduce foreign DNA into plant cells . Gold or tungsten particles (1–2 μm) are coated with the DNA to be used for transformation. The coated particles are loaded into a particle gun and accelerated to high speed either by the electrostatic energy released from a droplet of water exposed to high voltage or using pressurized helium gas; the target could be plant cell suspensions, callus cultures, or tissues. The projectiles penetrate the plant cell walls and membranes. As the microprojectiles enter the cells, transgenes are released from the particle surface for subsequent incorporation into the plant's chromosomal DNA. The microprojectile bombardment method was initially named as biolistics by its inventor Sanford (1988). Two types of plant tissue are commonly used for particle bombardment-Primary explants and the proliferating embryonic tissues.
5. **Transformation:** This method is used for introducing foreign DNA into bacterial cells e.g. *E.coli*. The transformation frequency (the fraction of cell population that can be transferred) is very good in this method. E.g. the uptake of plasmid DNA by E. coli is carried out in ice cold $CaCl_2$ (0-50C) followed by heat shock treatment at 37-45°C for about 90 sec. The transformation efficiency refers to the number of transformants per microgram of added DNA. The $CaCl_2$ breaks the cell wall at certain regions and binds the DNA to the cell surface.

6. **Conjunction:** It is a natural microbial recombination process and is used as a method for gene transfer. In conjunction, two live bacteria come together and the single stranded DNA is transferred via cytoplasmic bridges from the donor bacteria to the recipient bacteria.

7. **Lipofection:** Liposomes are circular lipid molecules with an aqueous interior that can carry nucleic acids. Liposomes encapsulate the DNA fragments and then adhere to the cell membranes and fuse with them to transfer DNA fragments. Thus, the DNA enters the cell and then to the nucleus. Lipofection is a very efficient technique used to transfer genes in bacterial, animal and plant cells.

Advantages of Transgenic Plants

1. Increased crop productivity.
2. Improvement in insect and disease resistance.
3. Improvements in food processing.
4. Improved nutritional value.
5. Better flavor.
6. Herbicides resistance.

A brief history of Transgenic Plants

- 1983- First transgenic tobacco plant resistant to kanamycin.
- 1985- Genetically engineered plants resistant to insects, viruses and bacteria were field tested for the first time.
- 1990- First successful field trial for *Bt* cotton was conducted.
- 1994- Flavr Savr Tomato designed for delayed ripening was approved for sale in USA.
- 1995-1996- Roundup ready soybean which was resistant to herbicides.
- Yield Gard Corn, which was protected from the corn borer, is approved for sale in the United States.
- Bollgard cotton first commercialized in the US.
- 1996- Herbicide tolerant Soybean was introduced.
- 1997- Roundup Ready cotton was first commercialized in the US.
- 1998- DEKALB markets the first Roundup Ready corn.
- 2006- Herbicide tolerant alfalfa was commercialized for first time in US.

24

Molecular Markers

Molecular markers, including DNA based genetic markers have found their applications in almost all areas of biology, more so in evolutionary biology, genetics and breeding in order to identify a particular sequence of DNA in a pool of unknown DNA. For many applications, DNA markers are preferred over other molecular markers since DNA is stable within an organism over time and among stages of development whereas other molecules are dynamic. A molecular marker is a DNA sequence that is readily detected and whose inheritance can be easily monitored. The uses of molecular markers are based on the naturally occurring DNA polymorphism. Most commonly used molecular markers in biological science include:

1. Restriction Fragment Length Polymorphism (RFLP)

RFLP is the first molecular marker technique which was invented by Botstein *et al.* (1980). The technique is based on restriction fragment length polymorphism; a molecular marker based on the differential hybridization of cloned DNA to DNA fragments in a sample of restriction enzyme digested DNAs; the marker is specific to a single clone/restriction enzyme combination. It consists of the following steps:

1. Isolation of differences in the DNA of the selected individuals.
2. Treat the genomic DNA with the specific restriction enzyme.
3. Separation of differnent sized DNA fragments on gel electrophoresis.
4. Hybridization with suitable probe.
5. Visualize them, by using X-rays film.

2. Random Amplified Polymorphic DNA (RAPD)

RAPD makers were initially described by William *et al.*, 1990. This molecular marker is based on the PCR amplification of random locations in the genome of the plant. With this technique, a single oligonucleotide is used to prime the amplification of genomic DNA. Because these primers are 10 nucleotides long, they have the possibility of annealing at a number of locations in the genome.

For amplification products to occur, the binding must be to inverted repeats sequences generally 150-4000 base pairs apart. The number of amplification products is directly related to the number and orientation of the sequences that are complementary to the primer in the genome. They are used for linkage mapping. RAPD includes following steps:

Isolation of DNA ——— Keeping the tubes in thermo cyclers ——— DNA strand separate (Denaturation the DNA at 94°C for 1 min.) ——— Annealing of primer (34°C for 2min.) ——— Primers annealed to the template DNA strand ——— Synthesis of DNA (72°C for 1.5 min). Separate amplified products by GEL.

3. Amplified Fragment Length Polymorphism (AFLP)

Amplified Fragment Length Polymorphism (AFLP) is the most recent class of molecular loci. These loci are generated using a procedure that combines restriction digestion and PCR amplification. The power of this procedure is to generate a large number of mappable loci with a single amplification. This will help to saturate a region of the genome rather quickly. The drawback is that procedure is a bit time consuming and requires the running of a DNA sequencing gel. AFLP can be performed by using very small DNA samples. It was developed by Zabeau and Vos in 1993. AFLP consists of the following steps:

Digestion of DNA with two restriction enzymes ——— Ligation of double stranded adapters to the end of restriction fragments ——— Pre-amplification of ligated products directed by primers complementary to adapter and restriction site sequences ——— DNA amplification of subsets of restriction fragments using selective AFLP primers ——— Labelling of amplified products ——— Labelled products are separated by electrophoresis——— Expose to X-ray film to visualize the AFLP fingerprint

4. Variable Number of Tandem Repeats (VNTR)

VNTR are also known as minisatellites .The term minisatellites was introduced by Jeffrey *et al.* (1985). These loci contain tandem repeats that vary in the number of repeat units between genotypes and are referred to as variable number of tandem repeat. It includes the following steps:

Extraction of DNA ——— Digestion of DNA with endonuclease ——— Separation of fragments by GEL ——— Transfer the fragments to a nylon filter paper ——— Hybridize with the minisatellite probes ——— visualization of polymorphic bands under autoradiography.

5. Single Nucleotide Polymorphism (SNP)

A single nucleotide polymorphism or SNP is an individual nucleotide base difference between two homologous DNA sequences. They are hailed as the newest markers of choice since they make up about 90% of all genetic variation. As a nucleotide base is the smallest unit of inheritance, SNPs provide the ultimate form of molecular marker. SNPs are meaningful only when their position is clearly defined and also with respect to a reference genome. Main features of SNP are presented below:

1. A novel class of DNA markers namely single nucleotide polymorphism in genome (SNPs) has recently become highly preferred in genomic studies.
2. The variations which are found at a single nucleotide position are referred to as SNP.
3. Most polymorphism of this type has two alleles.
4. Most common class of DNA polymorphism.
5. SNPs results due to substitution, deletion or insertion.
6. Small amount of DNA is required.

6. Sequence Tagged Sites (STS)

STS is a relatively short (200 to 500 bp), easily PCR-amplified sequence whose location in the genome is mapped. PCR amplification of STS marker is very specific and hence can be detected in the presence of all other genomic sequences. The STS concept was introduced by Olson *et al.* (1989). STS has following feature:

1. A sequence tagged site is a short DNA sequence that has a single copy in a genome and whose location and base sequence are known.
2. Usually 200-500 bp long
3. Detected by PCR
4. STSs are derived from cDNA.

7. Expressed Sequence Tags (ESTs)

ESTs are small pieces of DNA and their location and sequence on the chromosome are known. ESTs consists only exons. ESTs are type of STSs.

25

Marker Assisted Selection and Quantitative Trait Loci

Marker Assisted Selection (MAS)

Marker Assisted Selection [MAS] is defining as the indirect selection for a desired plant phenotype based on the banding pattern of linked molecular (DNA) markers. It is based to infer the presence of a gene from the presence of a marker which is tightly linked to the gene of interest and high heritability of the gene of interest. Marker assisted selection (MAS) is also known by other names as marker aided selection and marker assisted breeding (MAB).

Procedure of Marker Assisted Selection

The MAS consists of five important steps:

1. Selection of parents

Selection of suitable parents is first most important step in this technique. Parents chosen for MAS should have the contrasting characters or different origin. This will ensure identification of DNA of both the parents and also their segments in F_2 generation in various recombinations. The parents should be pure (homozygous).

2. Development of breeding population

The selected parents are crossed to obtained F_1 plants. F_1 plants are obtained by crossing two pure-lines or inbred lines, homogeneous ,but are heterozygous for all the RFLPs of two parents involved in the F_1. The F_2 progeny is required for the study of segregation pattern of RFLPs. Generally 50-100 F_2 plants are sufficient for the study of segregation of RFLP markers.

3. Isolation of DNA

DNA is isolated from the segregating population. DNA can be isolated even from the seedlings stage and no need to wait for flowering or seed development stage. The DNA is isolated from each plant of F_2 population. The isolated DNA is subjected to digestion with specific restriction enzyme to obtain fragments of DNA. The DNA fragments of different size are separated by subjecting the digested DNA to agarose gel electrophoresis. The gel is stained with ethidium bromide and the variation in DNA fragments can be viewed in the ultraviolet light.

4. Scoring RFLPs

The polymorphism in RFLPs between the parents and their recombinants in F_2 population is determined by using DNA probes. The labeled probes are used to find out the fragments having similarity. The probe will hybridize only with those segments which are complementary in nature.

5. Correlation with morphological traits

The DNA markers are correlated with morphological markers and the indirect selection through molecular markers is confirmed. Once the correlation of molecular markers is established with morphological markers, MAS can be effectively used for genetic improvement of various economic traits.

Applications of Marker Assisted Selection (MAS)

MAS have several useful applications in plant breeding, which have been summarized below:

1. Marker assistant selection is very effective, efficient and rapid method of transferring resistance to biotic and abiotic stresses in crop plants.
2. Useful in gene pyramiding for disease and insect resistance.
3. MAS are being used for transfer of male sterility and photo period insensitivity into cultivated genotypes from different sources.
4. MAS is being used for improvement of quality characters in different crops like, protein quality in maize, fatty acid (linolenic acid) content in soybean and storage quality in vegetables and fruit crops.
5. This selection scheme can be successfully used for transferring desirable transgene (such as Bt gene) from one cultivar to another.
6. Effective in introgression of desirable genes from wild into cultivated genotypes.
7. MAS is useful in genetic improvement of tree species .
8. MAS have wide application for genetic improvement of oligogenic traits as compared to polygenic traits.

Achievements of MAS

Crops such as maize, barley, rice, wheat, sorghum, soybean, chickpea, pea, sunflower, tomato, potato and some fruit crops has been improved by MAS. It has been mainly used for developing disease resistant cultivars in different crops. Some of the achievements are discussed here:

1. **Rice:** MAS has been successfully used for developing cultivars resistant to bacterial blight and blast in rice crop.
2. **Maize:** Normal lines in maize have been converted into quality protein maize (QPM) lines through MAS using opaque 2 recessive allele.
3. **Soybean:** In soybean, nematode resistant lines have been developed through MAS using SSR marker.
4. MAS have been used for genetic improvement of various characters in different crops such as, disease resistance, insect resistance, salinity resistance, shattering resistance etc.
5. Used for transfer of male sterility, photoperiod insensitivity, earliness, and improvement of protein content in some crop plants.
6. MAS are also being used for genetic improvement of forage and fruit crops.

Advantages of (MAS)

1. The accuracy of MAS, is very high, because molecular markers are not affected by environmental conditions
2. MAS take 3-5 years for developing a new cultivar.
3. MAS do not involve transgene.
4. MAS are effective for the genetic improvement of recessive characters.
5. Screening for traits such as root morphology and resistance to biotic and abiotic is very easy through MAS.
6. MAS are very effective method in accumulating multiple genes for resistance to specific pathogens and pests within the same cultivar. This process is called gene pyramiding.
7. MAS require only a small amount of plant tissue for DNA testing.
8. MAS permits mapping or tagging of quantitative trait loci (QTL) which is not possible by conventional method.

Disadvantage of MAS

1. MAS requires well equipped laboratory.
2. MAS require well trained manpower for handling of sophisticated equipments, isolation of DNA molecule and study of DNA markers.

3. The detection of various linked DNA markers (AFLP, RFLP, RAPD, SSR, SNP etc.) is a difficult, laborious and time consuming task.
4. MAS may become less efficient than phenotypic selection in the long term.
5. The use of MAS is more difficult for QTL because they have minor cumulative effects and are greatly influenced by environmental conditions and genetic background.

QTL (Quantitative Trait Loci)

In contrast to monogenic characteristics, such as disease and insect resistance, many important agronomic traits including yield, heading date, culm length, grain quality, and stress tolerance show continuous phenotypic variation. These complex traits usually are governed by a number of genes known as quantitative trait loci (QTLs) and these genes known as polygenes. The position of individual polygenes cannot be easily determined, which is main hindrance in cloning genes and production of transgenic plants for quantitative traits. Recently, however, methods have become available for studying the groups of polygenes i.e. called QTL mapping. The term was first coined by Gelderman (1975). A QTL is defined as "a region of the genome that is associated with an effect on a quantitative trait". Conceptually, a QTL can be a single gene, or it may be a cluster of linked genes that affect the trait.

QTL mapping studies have been reported in most crop plants for diverse traits including yield, quality, disease and insect resistance, abiotic stress tolerance, and environmental adaptation. QTL analysis is usually undertaken in segregating mapping populations, such as F_2-derived populations, recombinant inbred lines (RILs), near-isogenic lines (NILs), doubled haploid lines (DHs), and backcross populations.

Requirements for QTL Mapping

1. Suitable mapping population generated from phenotypically contrasting parents.
2. Saturated linkage map based on molecular markers.
3. Reliable phenotypic screening of mapping population.
4. Appropriate statistical packages to analyze the genotypic informations in combination with phenotypic informations for QTL detection.
5. The size of the mapping population for QTL analysis depends on several factors, including the type of mapping population employed for analysis, genetic nature of the target trait, objectives of the experiment, and the

resources available for handling a sizable mapping population in terms of phenotyping and genotyping.

6. While analysis of a large number of individuals (~500 or more) would enable detection of even QTLs having small effects on the target trait, from the practical point of view (MAS), the basic purpose of QTL mapping would be largely served if one can detect the QTLs with major effects. This would require, in general, a mapping population of a size of 200-300 individuals.

Methodology for QTL Mapping

1. Two inbred line that differ substantially in quantitative traits are crossed.
2. Segregating progeny, normally produced by backcross, are scored both for the quantitative traits and for a number of markers.
3. At each position on the genome defined by the markers, the most likely phenotypic effect of the putative QTL is computed together with odd ratio.
4. Based on the result of ratio it can be concluded whether QTL is located or not on predetermine position.

QTL Analysis Techniques

There are three basic techniques used in QTL analysis

1. Single point analysis-Uses one marker at a time
2. Flanking marker analysis-Uses a pair of markers simultaneously
3. Composite multipoint mapping-Uses multiple markers simultaneously.

Advantage of QTL

1. Provides a systematic and more reliable approach for dissecting polygenic variation.
2. Useful in determining the different breeding lines.
3. To enable positional cloning of the gene without QTL mapping.
4. QTL maps is very much useful in marker assisted selection.
5. Useful for tagging of simply inherited traits such as disease resistance.

Disadvantage of QTL

1. Require more number of markers.
2. Expensive techniques.
3. Mainly identifies loci with large effects.

4. Less strong ones can be hard to pursue.
5. Number of QTLs detected, their position and effects are subjected to statistical error.
6. Small additive effects/ epistatic loci are not detected and may require further analysis.

26

Intellectual Property Right (IPR)

Intellectual Property: Intellectual property is an intangible creation of the human mind, usually expressed or translated into a tangible form that is assigned certain rights of property. It includes, inventions, trademarks, industrial design and geographical indications.

Intellectual Property Rights: Intellectual property rights (IPR) are the rights given to people over the creation of their minds. They usually give the creator an exclusive right over the use of his/her creations for a certain period of time. It includes, creations of the mind: inventions, literary and artistic works, and symbols, names, images, and designs used in commerce.

Advantage of IPR

1. Provides exclusive rights to the creators or inventors.
2. Encourage individuals to distribute and share informations and ideas instead of keeping it confidential.
3. Provides legal defence and offers the creators with incentive for their work.
4. Help in social and financial development.

Laws of IPR in India

The Rules and Laws governing IPR in India are as follows:

1. The Copyright Act, 1957, The Copyright Rules, 1958 and International Copyright Order, 1999.
2. The Patents Act, 1970. The Patents Rules, 2003, The Intellectual Property Appellate Board (Patents Procedure) Rules, 2010 and The Patents (Appeals and Applications to the Intellectual Property Appellate Board) Rules, 2011.
3. The Trade Marks Act, 1999, The Trade Marks Rules, 2002, The Trade Marks (Applications and Appeals to the Intellectual Property Appellate Board) Rules, 2003 and The Intellectual Property Appellate Board (Procedure) Rules, 2003.

4. The Geographical Indications of Goods (Registration and Protection) Act, 1999 and The Geographical Indications of Goods (Registration and Protection) Rules, 2002.
5. The Designs Act, 2000 and The Designs Rules, 2001.
6. The Semiconductors Integrated Circuits Layout-Design Act, 2000 and The Semiconductors Integrated Circuits Layout-Design Rules, 2001.
7. The Protection of Plant varieties and Farmers' Rights Act, 2001 and The Protection of Plant varieties and Farmers Rights' Rules, 2003.
8. The Biological Diversity Act, 2002 and The Biological Diversity Rules, 2004.
9. Intellectual Property Rights (Imported Goods) Rules, 2007.

Duration of Intellectual Property Rights

1. Term of every patent is 20 years from the date of filing of patent application.
2. Term of every trademark registration is 10 years from the date of making of the application which is deemed to be the date of registration.
3. Copyright generally lasts for a period of sixty years.
4. The registration of a geographical indication is valid for a period of 10 years.
5. The duration of registration of Chip Layout Design is for a period of 10 years.
6. The duration of protection of registered varieties is different for different crops namely 18 years for trees and vines, 15 years for other crops and extant varieties.

Types of Intellectual Property Rights

The different types of Intellectual Property Rights are: **1.** Patents **2.** Copy rights **3.** Trademarks **4.** Trade secretes **5.** Industrial designs **6.** Geographical indications of goods **7.** Protection of Integrated Circuits layout design **8.** Biological diversity **9.** Plant varieties and farmers rights **10.** Undisclosed information.

1. Patent

The Patent System in India is governed by the Patents Act, 1970 as amended by the Patents (Amendment) Act, 2005 and the Patents Rules, 2003, as amended by the Patents (Amendment) Rules 2006 effective from 05-05-2006. Patent system in India is administered under the superintendence of the Controller General of Patents, Designs, Trademarks and Geographical Indications. The Head Office is located at Kolkata and other Patent Offices are located at Delhi, Mumbai and Chennai.

A patent is an exclusive right granted by law to applicants/ assignees to make use of and exploit their inventions for a limited period of time. Patent protection exists for twenty years from the date of filing of an application for utility and patents and fourteen years from the date of grant for design patents. After this period of time, the invention fall into the public domain and may be used by any person without permission.The patent holder has the legal right to exclude others from commercially exploiting his invention for the duration of this period. The invention to be patented must not be published in India or elsewhere, or in prior public knowledge or prior public use with in India or claimed before in any specification in India. Any invention concerning with composition, construction or manufacture of a substance, of an article or of an apparatus or an industrial type of process.

2. Copyright

The Copyright Act of 1957, The Copyright Rules, 1958 and the International Copyright Order, 1999 governs the copyright protection in India. It came into effect from January 1958. The Act has been amended in 1983, 1984, 1992, 1994 and 1999. Before the Act of 1957, copyright protection was governed by the Copyright Act of 1914 which was the extension of British Copyright Act, 1911.

Copy right refers to a document which grants exclusive right to the author/ creator to publish and sell literary or musical or artistic work. Creative expression may be captured in words, numbers, notes, sounds, pictures, or any other graphic or symbolic media. The owner of a copyright has the right to reproduce the work, prepare derivative works based on the original work (such as a sequel to the original), distribute copies of the work, and to perform and display the work. Copyright lasts for a certain time period after which the work is said to enter the public domain. Copy right includes, writings, paintings, musical works, dramatics works, audiovisual works, sound recordings, photographic works, broadcast, sculpture, drawings, architectural works etc.

Procedure for Registration

1. Application for registration is to be made on Form IV.
2. Separate applications should be made for registration of each work.
3. Each application should be accompanied by the requisite fee prescribed in the second schedule to the Rules.
4. The applications should be signed by the applicant or the advocate in whose favour a Power of Attorney has been executed.

3. Trademark

A trade mark is a visual symbol which may be a word signature, name, device, label, numerals or combination of colours used by one undertaking on goods or services or other articles of commerce to distinguish it from other similar goods or services originating from a different undertaking. Trade Marks are distinctive symbols, signs, logos that help consumer to distinguish between competing goods or services. The Trade, Marks Act, 1999 and the Trade Marks Rules, 2002 govern the law relating to Trade Marks in India. Example., trademark, service mark, collective mark and certification mark.

4. Trade secret

Trade secret consists of any valuable business information. The business secrets are not to be known by the competitor. There is no limit to the type of information that can be protected as trade secrets. Example: Recipes, marketing plans, financial projections, and methods of conducting business can all constitute trade secrets.

5. Design

A Design is the features of shape, configuration, pattern, ornamentation or composition of lines or colours applied to any article, whether in two or three dimensional (or both) forms. Design does not include any mode or principle of construction or anything which is mere mechanical device. It also does not include any trade mark or any artistic work. Designs are applied to a wide variety of products of different industries like handicrafts, medical instruments, watches, jewellery, house wares, electrical appliances, vehicles and architectural structures. The Designs Act, 2000 and the Designs Rules, 2001 presently govern the design law in India. The Act came into force on 25th May, 2000 while the Rules came into effect on 11th May 2001.

Requirements for Registration of Design

A design should be **1.** New or original **2.** Not be disclosed to the public **3.** Significantly distinguishable from known designs or combination of known designs **4.** Not comprise or contain scandalous or obscene matter **5.** Not be a mere mechanical contrivance. **6.** Be applied to an article and should appeal to the eye. **7.** Not be contrary to public order or morality.

Procedure of Registration

Any person who desires to register a design is required to submit the following documents to the Design Wing of the Patent Office at "Kolkata or any of the

Branch Offices of the Patent Office at Delhi, Mumbai and Chennai. The applications received by the Branch Offices will be transmitted to the Head Office for processing and prosecuting:

1. Application duly filed in on the prescribed form (Form-I) along with the prescribed fees, stating name in full, address, nationality, name of the article, class number, address for service in India. The application should also be signed either by the applicant or by his authorized agent.
2. Representation (in quadruplicate of size 33 cm x 20.5 cm with a suitable margin) of the article. Drawings / sketches should clearly show the features of the design from different views and state the view (e.g. front or side).
3. A statement of novelty and disclaimer (if any) in respect of mechanical action, trademark, work, letter, numerals should be endorsed on each representation sheet which should be duly signed and dated.
4. Power of attorney (if necessary).
5. Priority documents (if any) in case of convention application claimed under Section 44 of the Designs Act, 2000.

6. Geographical indications

Geographical indication referring to a country or to a place situated therein as being the country or place of origin of that product. Examples, Solapur Chaddar, Basmati Rice, Darjeeling Tea, Kanchipuram Silk Saree, Alphanso Mango, Nagpur Orange, Bikaneri bhujia etc. Laws of Geographical Indication of Goods (Registration and Protection) Act, 1999 and The Geographical Indications of Goods (Registration and Protection) Rules, 2002 deal with registration and better protection of geographical indications relating to goods.

Characteristics of Geographical Indication

1. It is an indication.
2. Originates from a definite geographical territory.
3. Used to identify agricultural, natural or manufactured goods.
5. The manufactured goods should be produced or processed or prepared in that territory.
6. Should have a special quality or reputation or other.

7. Semiconductor integrated circuit

Semiconductor integrated circuit is a product having transistors and other circuit elements, which are inseparably formed on a semiconductor material or an insulating material or inside the semiconductor material and designed to perform an electronic circuitry function. The layout-design of a semiconductor integrated circuit means a layout of transistors and other circuitry elements and includes lead wires connecting such elements and expressed in any manner in semiconductor integrated circuits. A period of 10 years counted from the date of filing an application for registration or from the date of first commercial exploitation anywhere in India.

8. Biological diversity

India has been a party to the Convention on Biological Diversity since 5th June, 1992 and ratified the Convention on 18th February, 1994. The Convention on Biological Diversity is one of the most broadly subscribed International Environmental Treaties in the world. It currently has 189 Parties - 188 States and the European Community - who have committed themselves to its three main goals: the conservation of biodiversity, sustainable use of its components and the equitable sharing of the benefits arising out of the utilization of genetic resources. The Secretariat of the Convention is located in Montreal, Canada. India is also a signatory to Cartagena Protocol on Biosafety signed on 23rd January 2001 and ratified on 11th September 2003.

Biodiversity Act, 2002 India enacted The Biological Diversity Act, 2002 and The Biological Diversity Rules, 2004 to fulfill its commitments in the Convention on Biological Diversity and in the Cartagena Protocol on Biosafety. The Biological diversity Act of 2002 contains 65 sections under 12 chapters while the Biological Diversity Rules of 2004 consists of 24 rules and one schedule. The Act provides for a three tier structure for regulation and access to Biological Diversity. The structure can broadly be summed up as :

1. **National Biodiversity Authority (NBA):** All matters relating to requests for access by foreign individuals, institutions or companies, and all matters relating to transfer of results of research to any foreigner is dealt with by the National Biodiversity Authority.
2. **State Biodiversity Boards (SBB):** All matters relating to access by Indians for commercial purposes will be under the purview of the State Biodiversity Boards (SBB). The Indian industry will be required to provide prior intimation to the concerned SBB about the use of biological resource. The State Board will have the power to restrict any such activity, which violates the objectives of conservation, sustainable use and equitable sharing of benefits.
3. **Biodiversity Management Committees (BMCs):** Institutions of local self government will be required to set up Biodiversity Management Committees in their respective areas for conservation, sustainable use, and documentation of biodiversity and chronicling of knowledge relating to biodiversity.

NBA and SBBs are required to consult the concerned BMCs on matters related to use of biological resources and associated knowledge within their jurisdiction.

9. Protection of Plant Varieties and Farmers' Right Act, 2001

The Protection of Plant Varieties and Farmers' Right Act, 2001 has been enacted to provide for the establishment of an effective system for protection of plant varieties, the rights of farmers and plant breeders and to encourage the development of new varieties of plants. The duration of protection of registered for trees and vines-18 years, other crops-15 years and for extant varieties-15 years from the date of notification of that variety by the Central Government under section 5 of the Seeds Act, 1966.

Farmers rights : Farmers have the rights to register varieties, benefit sharing, free services for registration, conducting tests on varieties, conservation of varieties, save, exchange, sell seeds, planting materials but he can not sell the seed of any brand name.

Plant breeders rights : This act confer exclusive right to the breeder or his successors, his agent or licensee to produce, sell, market, distribute, import or export seeds of his variety. Breeder has the right to authorize any other person for production or commercialization of his variety.

Objectives of the Act

1. To establish an effective system for protection of plant varieties, the rights of farmers and plant breeders and to encourage the development of new varieties of plants.
2. To recognize and protect the rights of the farmers in respect of their contribution made at any time in conserving, improving and making available plant genetic resources for the development of new plant varieties.
3. To protect plant breeders' rights to stimulate investment for research and development both in the public and private sector for development of new plant varieties.
4. To facilitate the growth of seed industry in the country that will ensure the availability of high quality seeds and planting material to the farmers.

Registration Procedure

The registration procedure consists of the following steps:

- Filling application in prescribed form.
- Payment of prescribed processing fee like Form charges Rs 200, DUS test fee Rs 20000, Annual fee Rs 2000, Registration charges Rs 5000, DUS Test site visit charges Rs 500.

- Initial scrutiny.
- Receipt.
- Acceptance after validation of information.
- Advertisement of application in Gazette of India.
- Invitations of objections within 3 months.
- Opposition notice with prescribed fee within 3 months.
- A copy of opposition notice referred by register to applicant.
- Counter statement by applicant within 2 months.
- Registrar refer a copy of counter statement to opposition and requiring submission of final.
- Registrar shall hear both the parties.
- After considering evidences, Registrar may uphold or reject opposition.
- Conduct DUS test (two years at two locations).
- Found distinct, uniform, stable and new.
- Issue of certification of registration.
- Publication of contents of certificate and invites claims of benefit sharing.
- Send a copy of claims to breeder of registered variety.
- Breeder may submit his notice of opposition to claim within three months.
- Sharing evidences given by both parties duly considered by the authority while disposing of any claim for benefit sharing.
- If claims are found to be genuine, breeder has to deposit amount of benefit.

Criteria for Registration of New Variety

1. **Novelty:** A new variety is granted protection, it it is deemed to be novel. It should not old more than one year. New variety should have a designation.
2. **Distinctiveness:** A new variety should be distinguishable by at least one essential characteristic from any other variety. The new variety may be different agronomic characters or polygenic traits or quality traits etc.
3. **Uniformity:** A new variety is deemed uniform if subject to the variation that may be expected from the particular features of its propagation, it is sufficiently uniform in its essential characteristics.
4. **Stability:** A new variety should give stable performance in different years or locations. Its essential characteristics remain unchanged after repeated propagation or, in case of a particular cycle of propagation, at the end of each such cycle.

The Agreement on Trade Related Aspects of Intellectual Property Rights (TRIPS)

The Agreement on Trade Related Aspects of Intellectual Property Rights (TRIPS) is an international agreement administered by the World Trade Organization (WTO) that sets down minimum standards for many forms of intellectual property (IP) regulation as applied to nationals of other WTO Members. The TRIPS agreement introduced intellectual property law into the international trading system for the first time and remains the most comprehensive international agreement on intellectual property to date. The following Intellectual Property Rights are covered under the TRIPS – Copyright, Geographical indications, Industrial designs, Integrated circuit layout-designs, Patents, New plant varieties, Trademarks and Undisclosed or Confidential information. TRIPS also specifies enforcement procedures, remedies, and dispute resolution procedures.

The World Intellectual Property Organization (WIPO)

The World Intellectual Property Organization (WIPO) is one of the 17 specialized agencies of the United Nations, located in Geneva, Switzerland. The Organization has External Offices in Brazil, Singapore and New York. The mission of WIPO is to promote innovation and creativity for the economic, social and cultural development of all countries, through a balanced and effective international intellectual property system. The two bureaus were united in 1893 and, in 1970, were replaced by the World Intellectual Property Organization, by virtue of the WIPO Convention. The WIPO Convention, the constituent instrument of the World Intellectual Property Organization (WIPO), was signed at Stockholm on July 14, 1967, entered into force in 1970 and was amended in 1979. WIPO is an inter-governmental organization that became in 1974 one of the specialized agencies of the United Nations system of organizations. WIPO currently has 185 Member States, and 68 intergovernmental organizations (IGOs) and 232 International non-governmental organizations (NGOs) and 63 National NGOs that are accredited as observers at WIPO meetings.

Features of WTO

1. Helps to promote piece.
2. Settling trade dispute.
3. Trade raises income.
4. Trade stimulates economic growth.
5. Make choice of product and quality.
6. Cooperating with other international organizations.
7. Assistance developing countries in trade police, issues etc.

Treaties for IPR protection

1. Berne Convention

The Berne Convention for the Protection of Literary and Artistic Works, usually known as the Berne Convention, is an International agreement governing copyright, which was first accepted in Bern, Switzerland in 1886. Berne Convention protects literary works, artistic works, dramatic works, musical works and cinematographic works and it also protects derivative works based on other pre-existing works, such as translation, adaptations, and arrangements of music and other alterations of a literary or artistic work. Berne Convention states the duration of the copyright protection as 50 years after the author's death. The Berne Convention was revised several times to cope up with the technological challenges that is, first revision took place in Berlin in 1908, followed by the revision in Rome in 1928, in Brussels in 1948, in Stockholm in 1967, and in Paris in 1971. The Convention rests on three basic principles and contains a series of provisions determining the minimum protection to be granted, as well as special provisions available to developing countries which want to make use of them.

2. Madrid Agreement

It concerning the international registration of marks came into force 1989 and was meant for repression of false or deceptive indications of source of goods.

Main features of Madrid Agreement

1. An applicant must be national member country. A person having his domicile or real and effective industrial or commercial interest in such country is also eligible.
2. A mark to be registered in member states should be first registered at the national level in the country of origin of the applicant.
3. The country having given the basic registration can only transmit the request for international filling to the international bureau of the WIPO along with the list of the courtier in which protection is being sought. There is no provision for directly filling a request under the agreement.
4. It is required that the country of origin has to be a member state. The role of the office of the country of origin is not only to send the application for International registration but also to certify that the mark, which is the subject of the international registration, is the mark, which has been registered in the country of origin.
5. For each application, a fee has to be paid for each designated country and WIPO.

3. Madrid Protocol

It relates to the Madrid Agreement concerning the international registration of marks entered in to force on December 1,1995 and came into operation on April1, 1996. Under this, a person can file as single trade mark application at the International office that will provide protection in multiple countries.

Features of Madrid Agreement

1. For the international registration, it is essential to first register marks at the national level. The time required for obtaining a mark at the national level. The time required for obtaining a mark at the national level varies from country to country.
2. With in one year designated member country has to examine and issue a notice of refusal by giving all the grounds for refusal.
3. A uniform fee designation for a member country. This was found in appropriate for countries with high level of national fees.
4. The only working language of Madrid Agreement is French.

4. Budapest Treaty

It is an international convention governing the recognition of microorganisms deposits in officially approved culture collections for the purpose of patent applications in any country that is a party to it. Because of the difficulties and on occasion of virtual impossibility of reproducing a microorganism from description of it in a patents pacification, it is essential to deposit a strain in culture collection centre for testing and examination by others. The treaty was signed in Budapest in 1973 and later amended in 1980. India has become a member of this treaty with effect from December 17, 2001.

27

Biometry

Biostatistics are the development and application of statistical methods to a wide range of topics in biology. It encompasses the design of biological experiments, the collection and analysis of data from those experiments and the interpretation of the results.

Quantitative traits are polygenic meaning they are controlled by many genes and gene interactions. Examples: height, weight, length etc.

Qualitative character

Qualitative traits are under the controlled of one or few genes, also called oligogeneic traits. For example, black or red coat color, horned or polled, coat color dilution, leaf color are all qualitative traits.

Biometrical Techniques

1. Genetic Variability Parameters

Variability is the basic requirement for improving the economic traits of any crop. The efficiency of selection largely depends upon the extent of genetic variability and high heritability of concerned character. Therefore, in crop improvement, only the genetic component is transmitted to the next generation. Estimation of variance components, coefficient of variation, broad sense heritability and genetic advance are used.

Genotypic variance (σ_g^2)

The genotypic variance is due to the differences among genotypes included in the study. Genotypic variances were calculated as per the formula suggested by Johnson *et al.* (1955).

$$\text{Genotypic variance}\,(\sigma_g^2) = \frac{\text{MSg} - \text{MSe}}{\text{r}}$$

Where,

MSg = Mean sum of square due to genotypes

MSe = Mean sum of square due to environment

r = Replications

Phenotypic variance (σ^2_p)

It is the sum of variance contributed by genetic causes and environmental factors. Phenotypic variances were calculated according to Comstock and Robinson (1952) .

Phenotypic variance (σ^2p) = σ^2g + σ^2e

Where,

σ^2g = Genotypic variance

σ^2e = Environmental variance

Error (Environmental) variance (σ^2)

The mean square of error represented by the variation attributed to environmental causes.

Environmental variance (σ^2) = MSe

Where,

MSe = Mean sum of square due to environment

Genotypic coefficient of variation (GCV)

$$GCV(\%) = \frac{\sqrt{\sigma^2 g}}{\overline{X}} \times 100$$

Phenotypic coefficient of variation (PCV)

$$PCV\ (\%) = \frac{\sqrt{\sigma^2{}_p}}{\overline{X}} \times 100$$

Where,

$\overline{X}$ = General mean

σ^2 g and $\sigma^2{}_p$ = Genotypic and phenotypic variances, respectively.

Heritability

Heritability refers to the portion of variability which is heritable in nature out of the total variability. Heritability is a good index of transmission of quantitative characters from parents to their offspring. Heritability and genetic advance are important selection parameters in crop improvement programme. It is due to the genetic causes. Heritability in broad sense was calculated according to the following formula as suggested by Burton and Devane (1953).

Heritability (h^2) =

Genetic advance (G.A.)

Improvement in the mean genotypic value (performance) of selected plants (progeny or lines) over the parental population (base population) is known as "Genetic Advance". Expected genetic advance represents the shift in a population mean towards superior side under selection pressure after single generation of selection. Characters having high heritability along with high GA indicate additive gene action. The expected genetic advance for each character was calculated according to the formulae suggested by Johnson *et al.* (1955).

G.A. = k x h^2 (b) x σp

Genetic advance as percentage of mean = $\frac{G.A.}{\overline{X}} \times 100$

Where,

h^2 = Heritability in broad sense $\frac{\sigma^2 g}{\sigma^2 p} \times 100$

k = Selection differential. It is a constant for a given selection intensity. It is taken as 2.06 at 5% selection intensity.

σp = Phenotypic standard deviation

$\overline{X}$ = General mean of the character

2. Correlation Coefficient

Correlation coefficients are used to measure the strength of the linear relationship between two variables. Correlation coefficients were suggested by Dewey and Lu (1959). Correlation is a statistical tool to find out the degree and direction of relationship between two or more variables. When change in one variable causes the change in another variable, the variables are said to be correlated. Correlation coefficient measure the degree of association, genotypic or phenotypic relationship between two or more characters which forms the basis for selection. Values always range between -1 to +1. Values at or close to zero imply a weak or no linear relationship. A value of exactly 1.0 means there is a perfect positive relationship between the two variables.

i) Genotypic correlation coefficients between character X and Y

$$r_{xy}(g) = \frac{COV_{xy(g)}}{\sqrt{V_{x(g)} \cdot V_{y(g)}}}$$

ii) Phenotypic correlation coefficients between character X and Y

$$r_{xy}(p) = \frac{COV_{xy(p)}}{V_{x(p)} \cdot V_{y(p)}}$$

3. Path Coefficient

Wright (1921) coined the term "path coefficient". To consider the relationships between variables in a multivariate system, a specific diagram-based approach was used. Estimating residual effect and determining yield determining characters. The analysis is based on linearity and additivity, and it also includes data on direct and indirect effects. The cause of association between two variables is measured using simple correlation.

Correlation studies along with path analysis provide a better way of understanding of the inter-relationship of different characters with grain yield which helps the breeder during selection.

4. Metroglyph Analysis

This analysis is used to assess pattern of morphological variation in a large number of genotypes used at a time. The pattern of variability is depicted by glyph on the graph. Semi-graphic method was developed by E Anderson in 1957.

D^2 statistic

P.C Mahalanobis (1928) developed this in anthropometry and psychometry and Rao (1952) suggested it for assessing genetic diversity in plants. A powerful method for determining genetic divergence. Genetic diversity is depicted in a cluster diagram.

5. Combining Ability

The concept of combining ability as a measure of gene action was proposed by Sprague and Tatum (1942) working on single cross in maize. Combining ability refers to ability of a genotype to transmit superior performance to its progenies. Combining ability analysis provides information about the nature and magnitude of various types of gene action involved in the expression of various quantitative characters. It also helps in identification of superior cross combinations. For estimating the combining ability analysis, crosses have to be attempted either in diallel or partial diallel or line x tester fashion, triallel, quadriallel, three way or double crosses. There are two types of combining ability viz, General combining ability and Specific combining ability.

General combining ability: The average performance of a strain or genotype in a series of hybrid combinations is termed as general combining ability. It is estimated from half sib families.

Specific combining ability: The performance of a parent in a specific combination is known as specific combining ability. SCA refers to the deviation of a particular cross from the general combining ability of its parents.

6. Methods of Combining Ability Analysis

Different methods (mating design) have been developed to estimate the general and specific combining ability by various workers.

- Diallel mating design (Griffing, 1956)
- Line x tester mating design (Kempthorne, 1957)
- Partial diallel analysis (Kempthorne and Curnow, 1961)
- Top cross (Davis, 1927)
- Polycross (Tysdal *et al.*)
- Triallel analysis (Rawling and Cockerham, 1962a)
- Quadriallel analysis (Rawlings and Cockerham, 1962b)

  Random mating designs
- Biparental cross (Comstock and Robinson, 1948 and expanded in 1952)
- Triple test cross (Kearsy and Jinks, 1968).

Diallel methods: Diallel methods (Griffing, 1956) have been extensively used in different species as it also provides information on *GCA* x ENV, *SCA* x ENV, maternal x ENV and non-maternal x ENV interactions, when applicable. It greatly improves researcher's efficiency in analyzing and interpreting diallel-cross data.

Line x tester analysis: A modified form of top cross which is used for measuring general and specific combining ability variances and used for evaluation of large number of germplasm lines at a time.

Partial diallel cross: A modified form of a diallel cross which utilizes only a part of all possible crosses for analysis form a diallel mating. Total number of crosses is equal to ns/2, where n and s are number of parents and sample crosses per parent.

Biparental Cross (North Carolina Design): Crossing of randomly selected plants from a random mating population in F_2 or subsequent generation of a cross between two pure lines in a definite fashion.

North carolina design-I (NCD-I)

A biparental cross in which each male is mated to a different group of females in F_2, resulting in f crosses, where f is the number of female plants used in a set for crossing. Nested Design is another name for this design.

North carolina design-II (NCD-II)

A biparental cross design in which each chosen male is mated to the same group of female plants in F_2, resulting in mf crosses, where m and f denote the number of male and female plants per set. Variance among single crosses is divided into three fractions: variance due to males, variance due to females, and variance due to males x females.

North carolina design-III (NCD-III)

A biparental cross design in which each randomly chosen male plant is mated to the P_1 and P_2 parents of the original cross in F_2, resulting in 2m crosses, where m is the number of male plants used per set. Variance among single crosses is divided into two fractions namely, variance among males and variance due to males x females.

7. Model of Generation Mean Analysis

There are three models of estimating gene effects and variances from generation means, viz.,

1. Six parameter model
2. Five parameter model and
3. Three parameter model

These models are based on the generations included in the study and presence and absence of epistasis.

Six parameter model

They involve six generations, namely P_1, P_2, F_1, F_2, B_1, and B_2. They estimate six parameters, namely m, d, h, i, j, and l. They require two crop seasons for material generation and a third crop season for testing. It is not possible to test this model with the (chi=square) X^2 test.

Five parameter model

This model was given by Hayman (1958). This model is used when the backcross progenies (B_1 and B_2) are not available and F_3 is available. P_1 P_2 F_1, F_2 and F_3. are the five generations used in the analysis. Five parameters are calculated: m, d, h, i and l. This model does not include information about epistasis of type j.

Three parameter model

This method was proposed by Jinks and Jones (1958). This is used in the absence of epistasis (non-allelic interactions). This model's analysis is based on six generations namely, P_1 P_2 F_1, F_2 B_1 and B_2. It calculates the values for three parameters that are m, d, and h. It takes two crop seasons to generate material and a third to test it. This does not include an (chi=square) X^2 test for model testing.

8. Stability Parameters

Phenotypically stable genotypes are of great importance because the environmental conditions vary from season to season and year to year. Genotype x environment is very impediment in plant breeding. Stability refers to the consistency in performance of a genotype over the environments. Eberhart and Russell (1966) suggested that the interaction of genotype with environment is a function of linear and non-linear response of genotype to the additive environmental variations. Thus, in addition to the linear component (regression coefficient) they included non-linear component, the "deviation from regression" as stability parameter.

Regression based models

- Finlay and Wilkinson (1963)
- Eberhart and Russell model (1966)
- Perkins and Jinks model (1968)
- Freeman and Perkins model (1971)

Stable Genotypes: The best genotypes are those with a higher mean and zero S^2. If bi = 1.0, it is appropriate for an unpredictable environment where bi> 1.0 is appropriate for a good environment and bi< 1.0 is appropriate for a poor environment.

Adaptability refers to the ability of genotypes to take advantage of the existing environmental conditions, whereas, stability relates to the ability of a genotype response in a highly predictable manner to an environmental condition. The existence of significant genotype x environment for grain yield that the analyzes of stability and adaptability were appropriate, the fact that the edaphoclimatic were the factors that most influence the adaptability and stability of genotypes.

28

Questions and Answers

A. Multiple Choice Questions

1. While studying the history of domestication of various cultivated plants were recognized earlier:
   i) Centers of origin ii) Centers of domestication
   iii) Centres of hybrid iv) Centres of variation
2. The quickest method of plant breeding is:
   i) Mutation breeding ii) Hybridization
   iii) Selection iv) Introduction
3. Polyploidy plants can arise spontaneously in nature by:
   i) Meiotic failure ii) Mitotic failure
   iii) Fusion of unreduced gametes iv) All of the above
4. Which is an allopolyploid:
   i) Canola ii) Cotton
   iii) Both a and b iv) Non of the above
5. Jagannath is a mutant variety of:
   i) Barley ii) Maize
   iii) Rice iv) Wheat
6. Autopolyploidy is found in:
   i) Coffee ii) Cotton
   iii) Both iv) None of the above
7. Cotton is:
   i) Cross pollinated ii) Never cross pollinated
   iii) Both iv) Often cross pollinated
8. Crosses between the plants of the same variety are called:
   i) Interspecific ii) Inter varietal
   iii) Inter generic iv) Intra varietal

9. Progeny obtained as a result of repeat self pollination of a cross pollinated crop is called:

   i) Pedigree line ii) Pure line

   iii) Inbreed line iv) Heterosis

10. Desired improved variety of economically useful crops are raised by:

   i) Natural Selection ii) Hybridization

   iii) Mutation iv) Biofertilisers

11. Primitive cultivars cultivated by farmers over generation and generation is known as:

   i) Modern cultivar ii) Obsolete cultivar

   iii) Land races iv) Mutants

12. How many center of origin were proposed by Vavilov initially:

   i) 6 ii) 8

   iii) 12 iv) 9

13. Seed collection that are disturbed only for regeneration are called as:

   i) Base Collection ii) Active Collection

   iii) Working Collection iv) Ficld Collcction

14. Protandry occurs in which of the following:

   i) Maize ii) Sugarbeet

   iii) Marigold iv) All of the Above

15. In case of barley, the chief quality characteristics is:

   i) Chaptti Making ii) Milling

   iii) Malting iv) Baking

16. Self-pollinated crops increases:

   i) Homozygosity ii) Homogenity

   iii) Heterozygosity iv) Heterogenity

17. Pigeon pea has chromosome number 2n=:

   i) 14 ii) 22

   iii) 28 iv) 56

18. Dominance hypothesis was given by :
    i) Mather ii) East and Shull
    iii) Jones iv) Davenport
19. In case of disease, score 0 indicates:
    i) Lack of disease symptom ii) Resistance
    iii) Lack of contact iv) Lack of Pathogen production
20. Term horizontal resistance was given by:
    i) Vander plank ii) Flor
    iii) Browning iv) Marshall
21. Polyploidy leads to rapid formation of new species through:
    i) Gene recombination
    ii) Development of multiple sets of chromosome
    iii) Isolation
    iv) Mutation
22. Seedless watermelon is a:
    i) Hexaploid ii) Tetraploid
    iii) Triploid iv) Pentaploid
23. Androgenic haploid culture was first performed by:
    i) Skoog and Miller ii) Steward
    iii) Halperrin iv) Guha and Maheshwari
24. Improved dwarf wheat variety with higher percentage of lysine and protein is:
    i) Kalyan Sona ii) Sharbati Sonara
    iii) Lerma Roja iv) Sonalika
25. Source of cytoplasmic male sterile gene in sorghum is:
    i) Tift 23 A ii) Milo
    iii) Kafir-60 iv) CMS-T
26. Bread wheat is a:
    i) Allopolyploidy ii) Autopolyploidy
    iii) Aneuploidy iv) Trisomic

27. Who coined the term heterosis:
    i) Transley ii) Robard
    iii) Shull iv) Hull
28. Aplant cell have the ability to develop in to complete plant is called:
    i) Tissue culture ii) Gene cloning
    iii) Genetic engineering iv) Totipotency
29. Which is the oldest breeding method:
    i) Introduction ii) Hybridization
    iii) Mutation breeding iv) Domestication
30. Hybridization involves:
    i) Removal of stigma ii) Removal of stamens
    iii) Male sterility iv) Self-incompatibility
31. Haploid from anther culture were first obtained in:
    i) Gossypium ii) Wheat
    iii) Datura iv) Nicotiana
32. Production of plants without fertilization is done by:
    i) Grafting ii) Transplanting
    iii) Vegetative propagation iv) Allthe above
33. Heterosis requires:
    i) Selection ii) Hybridization
    iii) Mutation iv) Introduction
34. Pure line variety refers to:
    i) Homozygosity ii) Heterozygosity
    iii) Heterosis iv) Heterogenious
35. Mating between more closely related individuals is called:
    i) Heterosis ii) Hybridization
    iii) Pure breeding iv) Inbreeding
36. Emasculation is a part of:
    i) Clonal selection ii) Male sterility
    iii) Hybridization iv) Mass selection

37. Plants having similar genotypes are called:
    i) Clone ii) Haploid
    iii) Autopolyploid iv) Genome
38. Bringing better varieties and plants from outside and acclimatizing them to local environment is called:
    i) Selection ii) Domestication
    iii) Centre of origin iv) Introduction
39. Gene responsible for dwarfing in wheat is:
    i) Bt 2 ii) Woogen
    iii) Norin 10 iv) Pal 1
40. Jaya and Ratna are the semi dwarf varieties of:
    i) Wheat ii) Rice
    iii) Gram iv) Maize
41. The species that are crossed to give sugarcane varieties with high sugar, high yield, thick stems and ability to grow in the sugarcane belt of North India:
    i) *Saccharum robustum* and *Saccharum officinarum*
    ii) *Saccharum barberi* and *Saccharum officinarum*
    iii) *Saccharum sinense* and *Saccharum officinarum*
    iv) *Saccharum barberi* and *Saccharum robustum*
42. Which of the following is type I restriction endonuclease:
    i) E co K ii) E co B
    iii) Alu I iv) E co K and E co B both
43. Which of the following is DNA modifying enzymes:
    i) Kinase ii) DNA Polymerase
    iii) Alkaline Phosphatase iv) All of above
44. Cloning large piece of DNA can be achieved by:
    i) Phase ii) Cosmids
    iii) YAC5 iv) All of above
45. Donald in 1968 proposed ideotype in which crop:
    i) Maize ii) Rice
    iii) Wheat iv) Sorghum

46. Single, homozygous self fertilized plant is called:
    i) Inbred ii) Isogenic line
    iii) Pureline iv) Multiline
47. Method used for development of oligogenic traits:
    i) Backcross ii) Bulk
    iii) Pedigree iv) Mass selection
48. Polycross test is most useful in which crop:
    i) Cleistogamous crops ii) Clonal Crops
    iii) Apomictic iv) Cross-pollinated crops
49. ICRISAT is situated at:
    i) Syria ii) Mexico
    iii) India iv) Philippines
50. When a cross is made between inbred and open pollinated variety is called:
    i) Single cross ii) Top cross
    iii) Multicross iv) Polycross
51. Triticale is:
    i) Monoploid ii) Diploid
    iii) Hexaploid iv) Triploid
52. Which of the following is not a part of hybridization:
    i) Tagging ii) Bagging
    iii) Emasculation iv) PCR
53. Bromato is made from:
    i) Brinjal ii) Tomato
    iii) Both i and ii iv) None of the above
54. Hybrid vigor is best maintained in:
    i) Hybridization ii) Emasculation
    iii) Vegetatively propagated crops iv) All the above
55. Heterosis lost of inbreeding is known as:
    i) Inbreeding depression ii) Out breeding
    iii) Hybrid vigor iv) Pure line

56. Crossing between two plants belonging to same variety is known as:
    i) Interspecific cross
    ii) Intervarietal cross
    iii) Intergenic cross
    iv) Intervarietal
57. The hybrid variety is:
    i) F_2
    ii) F_1
    iii) Male sterile
    iv) Inbreds
58. The methods of selection in plants are:
    i) Pure line selection
    ii) Mass selection
    iii) Hybridization
    iv) Both i and ii
59. Hybridization process performed between four inbreds, called:
    i) Single cross
    ii) Threeway cross
    iii) Double cross
    iv) Top cross
60. Heterosis refers as:
    i) Hybrid vigor
    ii) Inbreds
    iii) Isogenic lines
    iv) Inferiority of F_1
61. Co-dominant marker not used in MAS method of plant breading is:
    i) SSR
    ii) SNP
    iii) AFLP
    iv) All of the above
62. Ear to row method in maize was given by:
    i) Lonquist
    ii) Hopkins
    iii) Jones
    iv) Jenkins
63. RFLP-QTL linkage relationship can be studied in:
    i) Self pollinated species
    ii) Among inbred lines in cross-pollinated species
    iii) Self pollinated species and among inbred lines in cross-pollinated species both
    iv) None of these
64. Genes conferring insect resistance to plants are:
    i) Bt gene
    ii) Ipt gene
    iii) Pht gene
    iv) All of above

65. Source of Dwarfing gene in rice is found in the country:
    i) Japan ii) Taiwan
    iii) Philippines iv) India
66. The difference between population mean and mean of selected plants is known as:
    i) Heritability ii) Genetic Advance
    iii) Selection Differential iv) Direction al Selection
67. Which method is also known as evolutionary method of plant breeding:
    i) Bulk method ii) Backcross method
    iii) Pedigree method iv) Pureline method
68. Wheat variety MLKS 11 was developed by mixing seeds of how many near isogenic lines:
    i) 5 ii) 7
    iii) 8 iv) 9
69. Which is most widely used for rapid isolation of homozygous lines:
    i) Double haploid ii) SSD
    iii) Gynogenesis iv) Bulk method
70. Selection for simple inherited traits starts from which generation:
    i) F_2 ii) F_3
    iii) F_4 iv) F_5
71. Genetic variation in clones may arise due to:
    i) Segregation and recombination ii) Mechanical mixture
    iii) Somatic mutation iv) All of the above
72. The main aim of plant breeding is:
    i) To make soil fertile
    ii) To develop high yielding varieties
    iii) To control pollution
    iv) To become more progressive
73. Which of the following is not used for crop improvement:
    i) Hybridization ii) Mutations
    iii) Inbreeding iv) Introduction

74. A mane made allopolyploid cereal crop is:
    i) Gram ii) Raphano brassica
    iii) Triticale iv) *Avena sativa*
75. The world intellectual property organization was established in the year:
    i) 1886 ii) 1967
    iii) 1994 iv) 2001
76. Period of paten is valid for crops under PPV and FR Act, 2001 is:
    i) 5 years ii) 10 years
    iii) 15 years iv) 20 years
77. The head quarter of WTO is at:
    i) Geneva ii) USA
    iii) India iv) Africa
78. The uniformity is high in:
    i) Inbreds ii) Pure lines
    iii) F_1 hybrids iv) All the above
79. GI of goods, registration and protection act, 1999 was implemented from:
    i) 2000 ii) 2003
    iii) 1997 iv) 2014
80. A variety which has been cultivated and evolved by farmers in their fields is called:
    i) Extant variety ii) New variety
    iii) Farmers variety iv) Pure line
81. Types of area covered by TK:
    i) Agriculture ii) Horticulture
    iii) Forestry iv) All the above
82. What is not a design:
    i) Any trade mark ii) Pattern
    iii) Colour combination iv) Composition of lines
83. The Berne convention was adopted in:
    i) 1866 ii) 1872
    iii) 1886 iv) 1999

84. ITGRFA finally entered in to the force in the year:

i) 2001 ii) 2004

iii) 2006 iv) 2008

85. Madrid Agreement was concluded in:

i) 1880 ii) 1888

iii) 1890 iv) 1891

86. The registration for GI is valid for:

i) 5 years ii) 10 years

iii) 15 years iv) 20 years

87. Term nobilisation is related to which crop:

i) Sugarbeet ii) Potato

iii) Ginger iv) Sugarcane

88. Gametophytic SI was first described by:

i) Koelreuter ii) Hughes

iii) Eas iv) East and Mangelsdorf

89. The concept of biparental mating was originally developed by:

i) Comstock and Robinson ii) Hayman

iii) Russel iv) Eberhart

90. The term "Heterosis" was coined by:

i) Davenport ii) Jones

iii) Shull iv) East

91. Heterobeltiosis is also refer as:

i) Better parent heterosis ii) Commercial heterosis

iii) Average heterosis iv) Standard heterosis

92. Grid method was developed by :

i) Nillson ii) Gardner

iii) Jones iv) Johanseen

93. Which of the following the most important stage for disease escape:

i) Reproduction ii) Infection

iii) Establishment iv) Contact

94. Which plant part is used for mutagen treatment:

i) Pollen grain ii) Seed

iii) Vegetative propagule iv) All of the above

95. Totipotency is a feature of which of the following:
    i) Plant cell ii) Animal cell
    iii) Cancerous cell iv) All of the Above

96. Dwarf wheat varieties were developed by:
    i) Swaminathan ii) Vavilov
    iii) Borlaug iv) Shull

97. The product of hybridization is known as:
    i) Heterozygous ii) Homozygous
    iii) Clone iv) Hybrid

98. The basis of green revolution was:
    i) Plant breeding ii) Hybrid varieties
    iii) Extensive cultivation iv) Synthetic varieties

99. Plants can be disease resistant by:
    i) Hormone treatment
    ii) Breeding with their wild relatives
    iii) Pestiscides
    iv) Insecticides

100. The offspring from a cross between two individuals differing for atleast one character is called:
    i) Hybrid ii) Variant
    iii) Polyploidy iv) Hybrid

101. Which of the following is not ionizing in nature:
    i) X ray ii) Fast neutrons
    iii) UV ray iv) Gamma ray

102. Pigeon pea is a:
    i) OCP ii) CP
    iii) SP iv) None of these

103. Which of the year is declared as National Year of Millets in India:
    i) 2016 ii) 2018
    iii) 2019 iv) 2020

104. Triticale is a classical example of which hybridization:
   i) Inter species hybridization ii) Intra species hybridization
   iii) Inter generic hybridization iv) All the above

105. Gene for gene relation between a host and its pathogen was postulated by Flor based on his work on:
   i) Cowpea ii) Pea
   iii) Lathyrus iv) Linseed

106. Maize is believed to be originated from:
   i) India ii) America
   iii) Peru iv) China

107. Which of the following crops shows high degree of inbreeding depression:
   i) Carrot ii) Alfalfa
   iii) Both i and ii iv) Onion

108. Which of the following breeding methods is most commonly used for selection from segregating generations of crosses of self pollinated crops:
   i) Pedigree method ii) Bulk method
   iii) Backcross method iv) Pure line method

109. Self incompatibility is found in:
   i) Wheat ii) Mustard
   iii) Maize iv) Rice

110. Selection is not practiced upto F_5 or F_6 generation in which method:
   i) SSD method ii) Back cross method
   iii) Bulk method iv) Grid method

111. 2n-1 is the symbol of :
   i) Nullisomic ii) Trisomic
   iii) Tetrasomic iv) Monosomic

112. Hughes and babcock (1950) first described:
   i) Sporophytic self-incompatibility
   ii) Gametophytic self-incompatibility
   iii) Male sterility
   iv) Hetero morphic self-incompatibility

113. Collection of germplasm from foreign country is called as:
   i) Indigenous collection ii) Indirect collection
   iii) Direct selection iv) Exotic collection

114. In recurrent selection for specific combining ability, the tester may be :
   i) Open pollinated variety   ii) Inbred
   iii) Synthetic variety   iv) Clone
115. Normally IET is conducted for:
   i) One year   ii) Two year
   iii) Three year   iv) Four year
116. Vilmorin successfully used progeny test in which crop:
   i) Wheat   ii) French bean
   iii) Sugarbeet   iv) Sugarcane
117. Eat to row method was extensively used in:
   i) Pearlmillet   ii) Cotton
   iii) rice   iv) Maize
118. Breeding for disease resistance requires:
   i) A good source of resistance   ii) Planned hybridization
   iii) Diseases test   iv) All of these
119. Polyploidy is induced through:
   i) Colchicine   ii) Mutagenic chemicals
   iii) Ethylene   iv) Irradiation
120. Heterosis is:
   i) Appearance of spontaneous mutations
   ii) Induction of mutations
   iii) Superiority of hybrids over their parents
   iv) Mixture of two or more traits
121. Somatic hybridization is achieved through:
   i) Protoplast fusion
   ii) Conjugation
   iii) Grafting
   iv) Recombinant DNA technology
122. The most viable component in the medium is:
   i) Carbon source   ii) Growth regulator
   iii) Inorganic elements   iv) Vitamins

123. Which of these molecular markers is not based on PCR:

i) RAPD ii) AFLP

iii) RFLP iv) SSR

124. Notification of released variety is issued by:

i) NSC ii) SSC

iii) SFCI iv) Govt.of India

125. Hybrid seed can be produced through open pollination with the use of:

i) Inbreds ii) Clones

iii) Pure lines iv) All the above

126. Bt genes are used for transgeneic plants having:

i) Virus resistance ii) Bacterial resistance

iii) Insect reistance iv) Antibiotic resistance

127. Bagging is done to:

i) Avoid male sterility

ii) Avoid self pollination

ii) Achieve desired pollination

iv) Prevent contamination from foreign pollen

128. A technique of micropropagation is:

i) Multiple root production

ii) Somatic embryogenesis

iii) Growth of micro organisms on culture medium

iv) Multiple shoot production and embryo rescue

129. Golden rice is rich in:

i) Vitamin C ii) Protein

iii) B-carotene iv) Antioxidant

130. For corrections of defects in any popular variety which method is used :

i) Bulk method ii) Backcross method

iii) Pedigree method iv) Pureline method

131. Sub lethal mutation kills:

i) Less than 50% individuals ii) More than 50% individuals

iii) 100% individuals iv) Donot kills the individuals

132. The first step in genetic improvement of plant species is :

i) Germplasm collection ii) Hybridization

iii) Selection iv) Domestication

133. Two heterozygous testers are used in:
   i) Reciprocal recurrent selection
   ii) Recurrent selection for GCA
   iii) Recurrent selection for SCA
   iv) Simple recurrent selection
134. Which crop is more salt tolerant:
   i) Maize ii) Sorghum
   iii) Barley iv) Gram
135. The lines which are identical in all respect,except for one gene is a:
   i) Puelines ii) Inbred lines
   iii) Hybrids iv) Isogenic lines
136. After 2-3 weeks, the meristem cultures are transferred to:
   i) Browning medium ii) Root multiplication medium
   iii) Shoot multiplication medium iv) Leaf multiplication medium
137. The term vertical and horizontal resistance were introduced by:
   i) Flor ii) Vander plank
   iii) Cramer iv) White
138. Best plant for protoplast isolation is:
   i) Meristem ii) Root
   iii) Shoot tip iv) Leaf
139. When embryo develops from antipodal or synergid cells, it is called:
   i) Parthenogenesis ii) Apogamy
   iii) Apospory iv) Diplospory
140. Over-dominance hypothesis of heterosis was proposed by:
   i) Davenport ii) Shull
   iii) East iv) Both (ii) and (iii)
141. In three line system of male sterility,which of the following line is not required:
   i) A-line ii) B-line
   iii) R-line iv) C-line
142. Wheat has chromosome number:
   i) 2n=36 ii) 2n=52
   iii) 2n=28 iv) 2n=42

143. Which one is a chemical mutagen:
   i) Gamma rays
   ii) Base analogue
   iii) X-rays
   iv) Gamma rays
144. Inbreeding depression is observed in:
   i) F_1
   ii) F_2
   iii) Bc_1
   iv) F_3
145. Quantitative characters are governed by:
   i) Few genes
   ii) Poly genes
   iii) High heritability
   iv) Discontinuous variation
146. When F_1 is crossed with one of its recessive parents, it is known as:
   i) Double cross
   ii) Top cross
   iii) Back cross
   iv) Test cross
147. Which breeding method is most commonly used for development of disease resistant varieties:
   i) Divergent method
   ii) Bulk method
   iii) Back cross method
   iv) Mass selection
148. Best method for population improvements is:
   i) Heterosis breeding
   ii) Mass selection
   iii) Recurrent selection
   iv) Synthetic breeding
149. Asexual reproduction occurs in:
   i) Sugarcane
   ii) Potato
   iii) Pearlmillet
   iv) Both i and ii
150. Most commonly used culture medium in tissue culture is:
   i) White medium
   ii) Gomberg's medium
   iii) Perk' s medium
   iv) M.S. medium
151. Self pollination is a feature of:
   i) Unisexual crops
   ii) Bisexual crops
   iii) Asexual crops
   iv) Clonal crops
152. Heterosis can be maintained for longer period by:
   i) Hybridization
   ii) Mutation
   iii) Tissue culture
   iv) Clonal selection

153. In double trisomic, the chromosome constitution would be :

i) 2n+2 ii) 2n+1

iii) 2n+1+1 iv) 2n-1-1

154. RMO 257 is a popular variety of:

i) Mungbean ii) Urdbean

iii) Cowpea iv) Mothbean

155. Diploid apogamy occurs in:

i) Antennaria ii) Alchemilla

iii) Allium iv) All of the above

156. Isogenic lines are developed by:

i) Pedigree method ii) Back cross method

iii) SSD method iv) Bulk method

157. In tetra somic, the chromosome constitution is represented by:

i) 2n+1 ii) 2n+2

iii) 2n+1+ iv) 2n-1-1

158. Mutagenic action of X-rays on Drosophilla was discovered by:

i) Muller ii) Stadlar

iii) Vries iv) Roentgen

159. In recurrent selection for specific combining ability, tester may be :

i) Open pollinated variety ii) Synthetic variety

iii) Inbred iv) Composite variety

160. Clones are highly:

i) Homozygous ii) Heterozygous

iii) Stable iv) Homoheterozygenous

161. Genetic variability can be created by:

i) Hybridization ii) Mutation

iii) Poly ploidy iv) Allthe above

162. Dee-Gee-Woo-Gen is found in:

i) Rice ii) Wheat

iii) Gram iv) Maize

163. Components of genetic resources are:

i) Land race ii) Obsolete cultivar

iii) Wild relatives iv) Allthe above

164. 2n-2 is the symbol of:
   i) Monosomic ii) Nullisomic
   iii) Trisomic iv) Tetrasomic
165. The important attributes of drought escape is :
   i) Broad leaves ii) Late maturity
   iii) Early maturity iv) More tillers
166. Qualitative characters are governed by:
   i) Few genes ii) Poly genes
   iii) Chromosomal genes iv) Allthe above
167. The case,where flowers do not open at all, is known as:
   i) Dicogamy ii) Cleistogamy
   iii) Chasmogamy iv) Geitonogamy
168. Pure line theory was proposed by:
   i) Hull ii) Strasburger
   iii) Gardner iv) Johannsen
169. Source of irradiation in gamma garden is:
   i) Zinc ii) Protons
   iii) Cobalt-60 iv) Helium
170. Central Rice Research Institute is located at:
   i) Maxico ii) Kolcutta
   iii) Philippines iv) Cuttack
171. Monosomic have been reported in:
   i) Wheat ii) Tobacco
   iii) Oat iv) All the above
172. Which of the following is not a example of half sib family selection:
   i) S_1 family selection ii) Ear to row
   iii) RSSCA iv) RSGCA
173. Plant breeding is a:
   i) An art ii) A science
   iii) both i and ii iv) None of the above
174. Clones are degenerated due to:
   i) Viral infection ii) Bacterial infection
   iii) Mutation iv) All the above

175. Highest yield potential is in:
   i) Single cross hybrid
   ii) Double cross hybrid
   iii) Three way cross hybrid
   iv) Synthetic varieties
176. TGMS and PGMS are being used for hybrid seed production in:
   i) Pearlmillet
   ii) Rice
   iii) Maize
   iv) Barley
177. Transgressive is done to obtained:
   i) Normal segregants
   ii) Segregants like parents
   iii) Segregants exceeding both the parents
   iv) Heterozygous
178. RSSCA was first proposed by:
   i) Hayes
   ii) Hull
   iii) Garber
   iv) Jones
179. Classification of self incompatibility was suggested by:
   i) Koelreuter
   ii) Gardner
   iii) East
   iv) Lewis
180. In mass selection ,plants are selected on the basis of :
   i) Phenotype
   ii) Genotype
   iii) Progeny test
   iv) None of the above
181. A character can be easily transferred by back cross method when it has:
   i) Low heritability
   ii) Medium heritability
   iii) High heritability
   iv) All the above
182. The homozygous and heterozygous words were first used by:
   i) Bateson
   ii) Darwin
   iii) Grifth
   iv) Mendel
183. Crossing without emasculation is done in:
   i) Cucurbits
   ii) Maize
   iii) Bajara
   iv) All the above
184. Early testing was proposed by:
   i) Jenkins
   ii) Davis
   iii) Shull
   iv) Hull

185. B line has:
   i) Genetic male sterility ii) Maintainer line
   iii) Cytoplasmic male sterile line iv) Restorer gene

186. The off season crop of maize may be grown at:
   i) Cuttack ii) Delhi
   iii) Wellington iv) Dholi

187. The individual plant selection was developed in detailed by:
   i) Vilmorin ii) Mendel
   iii) Nilsson-Ehle iv) Johannsen

188. The concept of centre of origin first given by:
   i) Knight ii) Vavilov
   iii) Vilmorin iv) Cooper

189. In sporophytic system of incompatibility, the mating between S_1 S_2 and S_2 S_3 Would produce:
   i) Fully incompatible ii) Fully compatible
   iii) Partially compatible iv) All the above

190. Haploid plants can be obtained from:
   i) Anther culture ii) Pollen culture
   iii) Ovary culture iv) All the above

191. X- rays were first discovered by:
   i) Muller ii) Roentgen
   iii) Mullert iv) Stadler

192. R line have:
   i) Fertility restoring genes ii) Male sterility
   iii) Male fertility iv) None of the above

   Ans. **i**

193. Individuals having single set of chromosome in a cell is called:
   i) Diploid ii) Tetraploid
   iii) Monoploid iv) Triploid

194. Protoandry condition is found in:
   i) Pearlmillet ii) Isabgol
   iii) Wheat iv) Maize

195. Triticale was produced by crossing:
   i) What and rye
   ii) Wheat and oat
   iii) Wheat and barley
   iv) Wheat and rice
196. EMS is a:
   i) Acridines
   ii) Base analogues
   iii) Alkylating agents
   iv) Physical agents
197. The best method for producing virus-free plants in sugarcane is:
   i) Meristem culture
   ii) Apical bud culture
   iii) Organ culture
   iv) Axillary bud culture
198. Variety bio-902 was developed through:
   i) Mutation breeding
   ii) Back cross method
   iii) Pedigree method
   iv) Tissue culture
199. Tissue culture media is sterilized by:
   i) Dry sterilization
   ii) Filter sterilization
   iii) Autoclaving
   iv) Flame sterilization
200. Who is the father of the Green revolution in India:
   i) M.S. Swaminathan
   ii) Charles Darwin
   iii) Herbert Boyer
   iv) Stanley Cohen
201. Which phase is often referred to as the Green Revolution:
   i) The Mid-1900s
   ii) The Mid-1980s
   iii) The early2000s
   iv) The Mid-1960s
202. Which of the following is not related to plant breeding:
   i) Helped to increase the yield of crops
   ii) Purpose ful manipulation of plant species
   iii) Gives disease-resistant plants
   iv) Not suited for cultivation
203. Which of the following is not a trait that should be incorporated in a crop plant:
   i) Decreased tolerance to environmental stresses
   ii) Increased yield
   iii) Resistance to pathogens
   iv) Increased tolerance to insect pests

204. Which of the following is not the main step in carrying out plant breeding technique:
   i) Collection of variability
   ii) Cross among the diseased parents
   iii) Selection and testing of superior recombinants
   iv) Evaluation and selection of parents

205. Which of the following step is the main root of any plant breeding programme:
   i) Genetic variability
   ii) Evaluation and selection of parents
   iii) Cross among these selected parents
   iv) Selection and testing of superior recombinants

206. Which of the following is not included in germplasm collection:
   i) Wild relatives
   ii) Old improved varieties
   iii) Diseased varieties
   iv) Pure lines

207. Restriction enzymes were discovered by:
   i) Smith and Nathans
   ii) Alexander Fleming
   iii) Berg
   iv) None

208. Bacteria protect themselves from viruses by fragmenting viral DNA with:
   i) Ligase
   ii) Endonuclease
   iii) Exonuclease
   iv) Gyrase

209. Southern blotting is:
   i) Attachment of probes to DNA fragments
   ii) Transfer of DNA fragments from electrophoresis gel to a nitrocellulose sheet
   iii) Comparison of DNA fragments to two sources
   iv) Transfer of DNA fragments to electrophoresis gel from cellulose membrane

210. The Golden Rice variety is rich in:
   i) Vitamin C
   ii) β-carotene
   iii) Biotin
   iv) Lysine

211. The DNA fragments have sticky ends due to:
   i) Endonuclease
   ii) Unpaired bases
   iii) Calcium ions
   iv) Free methylation

212. Plasmids are used as cloning vectors for which of the following reasons:
    i) Can be multiplied in culture
    ii) Self-replication in bacterial cells
    iii) Can be multiplied in laboratories with the help of enzymes
    iv) Replicate freely outside bacterial cells

213. Which is a genetically modified crop:
    i) Bt-cotton ii) Bt-brinjal
    iii) Golden rice iv) All

214. PCR technique was invented by:
    i) Karry Mullis ii) Boyer
    iii) Sanger iv) Cohn

215. The first transgenic plant to be produced is:
    i) Brinjal ii) Tobacco
    iii) Rice iv) Cotton

216. Which of the following is the quality of improved transgenic basmati rice:
    i) Gives high yield but no characteristic aroma
    ii) Gives high yield and is rich in vitamin A
    iii) Does not require chemical fertilizers and growth hormones
    iv) Resistant to insects and diseases

217. Excision and insertion of a gene is called
    i) Biotechnology ii) Genetic engineering
    iii) Cytogenetic iv) Gene therapy

218. The expression of a transgene in the target tissue is identified by a:
    i) Transgene ii) Promoter
    iii) Enhancer iv) Reporter

219. Intellectual Property Rights (IPR) protect the use of information and ideas that are of:
    i) Ethical value ii) Moral value
    iii) Social value iv) Commercial value

220. The term 'Intellectual Property Rights' covers:
    i) Copyrights ii) Know-how
    iii) Trade dress iv) All of the above

221. The following can not be exploited by assigning or by licensing the rights to others:

i) Patents
ii) Designs
iii) Trademark
iv) All of the above

222. The following can be patented:

i) Machine
ii) Process
iii) Composition of matter
iv) All of the above

223. Trade mark:

i) Is represented graphically
ii) Is capable of distinguishing the goods or services of one person from those of others
iii) May include shapes of goods or combination of colors
iv) All of the above

224. Symbol of Maharaja of Air India is:

i) Copyright
ii) Patent
iii) Trademark
iv) All of the above

225. In India, the literary work is protected until:

i) Lifetime of author
ii) 25 years after the death of author
iii) 40 years after the death of author
iv) 60 years after the death of author

226. Design does not include:

i) Features of shape
ii) Composition of lines or colors
iii) Mode or principle of construction
iv) None of the above

227. Which of the following is (are) included in Geographical indications of Goods:

i) Handicraft
ii) Foodstuff
iii) Manufactured
iv) All of the above

228. Patent is a form of:

i) Tangible Property
ii) Intellectual Property
iii) Industrial property
iv) Both (ii) and (iii)

229. Patent protects:

i) Discovery ii) Invention

iii) New invention iv) Both (i) and (ii)

230. Invention means:

i) New product having inventive step and capable industrial application

ii) New process

iii) New product or process having inventive step and capable industrial application

iv) None of the above

231. Patent right is:

i) Exclusive right ii) Natural right

iii) Property right iv) Both (i) and (iii)

232. Patentability criteria includes:

i) Novelty ii) Inventive step

iii) Capable of Industrial application iv) All the above

233. Prior art search includes:

i) Search of Patent literatures

ii) Search of Non-patent literature

iii) Both (i) and (ii)

iv) None of the above

234. IPC means:

i) Indian Patent Classification

ii) International Panel Code

iii) International Patent Classification

iv) International Postal Code

235. Which of the following is not a category of copyright work in the CDPA:

i) Literary works ii) Furniture

iii) Sculpture iv) Musical work

236. Which of the following is not an example of a literary work:

i) A character from a novel ii) A shopping list

iii) A textbook iv) A computer programme

237. Which of the following statements is correct regarding artistic works:

i) All forms of art are protected as artistic works

ii) Artistic quality is irrelevant where copyright protection is concerned

iii) Only works falling within the definition of artistic works in s. 4 CDPA are protected

iv) The question of whether something is art or not is judged objectively

238. The quickest method of plant breeding is:

i) Introduction | ii) Selection
iii) Hybridization | iv) Mutation breeding

239. Plants having similar genotypes produced by plant breeding are called:

i) Clone | ii) Haploid
iii) Autoploid | iv) Genome

Answers

1	i	29	i	57	ii	85	iv	113	iv	141	iv	169	iii
2	iv	30	ii	58	iv	86	ii	114	ii	142	iv	170	iv
3	iv	31	iii	59	iii	87	iv	115	i	143	ii	171	iv
4	ii	32	iii	60	i	88	iv	116	iii	144	ii	172	i
5	iii	33	ii	61	iii	89	i	117	iv	145	ii	173	iii
6	i	34	i	62	ii	90	iii	118	iv	146	iv	174	iv
7	iv	35	iv	63	iii	91	i	119	i	147	iii	175	i
8	iv	36	iii	64	iv	92	ii	120	iii	148	iii	176	ii
9	iii	37	i	65	ii	93	iv	121	i	149	iv	177	iii
10	ii	38	iv	66	ii	94	iv	122	ii	150	iv	178	ii
11	iii	39	iii	67	i	95	i	123	iii	151	ii	179	iv
12	ii	40	ii	68	iii	96	iii	124	iv	152	iv	180	i
13	i	41	ii	69	i	97	iv	125	i	153	iii	181	iii
14	iv	42	iv	70	i	98	ii	126	iii	154	iv	182	i
15	iii	43	iv	71	iv	99	ii	127	iv	155	iv	183	iv
16	i	44	iv	72	ii	100	i	128	ii	156	ii	184	i
17	ii	45	iii	73	iii	101	iii	129	iii	157	ii	185	ii
18	iv	46	iii	74	iii	102	i	130	ii	158	i	186	iv
19	ii	47	i	75	ii	103	ii	131	ii	159	iii	187	iii
20	i	48	iv	76	iii	104	iii	132	iv	160	ii	188	ii
21	ii	49	iii	77	i	105	iv	133	i	161	iv	189	ii
22	iii	50	ii	78	iv	106	ii	134	iii	162	i	190	iv
23	iv	51	iii	79	ii	107	iii	135	iv	163	iv	191	ii
24	ii	52	iv	80	iii	108	i	136	iii	164	ii	192	i
25	iii	53	iii	81	iv	109	ii	137	ii	165	iii	193	iii
26	i	54	iii	82	i	110	iii	138	iv	166	i	194	iv
27	iii	55	i	83	iii	111	iv	139	ii	167	ii	195	i
28	iv	56	iv	84	ii	112	i	140	iv	168	iv	196	iii

197	i	204	ii	211	ii	218	iv	225	iv	232	iv	239	i
198	iv	205	i	212	ii	219	iv	226	iii	233	iii		
199	iii	206	iii	213	iv	220	iv	227	iv	234	iii		
200	i	207	i	214	i	221	iii	228	iv	235	ii		
201	iv	208	ii	215	ii	222	iv	229	iii	236	iv		
202	iv	209	iv	216	ii	223	iv	230	iii	237	i		
203	i	210	ii	217	ii	224	iii	231	iv	238	ii		

B. Fill in the Blanks

1. 2n-2 is the symbol of ________ **Nullisomic** ________ .
2. Gene-for-gene hypothesis was given by Flor in Flax rust caused by organism ________ ***Melampsora lini***.
3. Transfer of high sugar content from Tropial cane to Indian cane is termed as ________ **Nobilization** ________ .
4. In ________ **Sporophytic** ________ system of self incompatibility, the recovery of both male and female is possible.
5. The phenomenon in which single oligo gene affects the expression of more than one character is known as ________ **Pleiotropic** ________.
6. International Crops Research Institute for Semi-Arid Tropics is situated at ________ **Hyderabad**________ .
7. ________ **Inbred lines** ________ is used as a tester in recurrent selection for specific combining ability.
8. Clones are highly ________ **heterozygous** ________ .
9. Additive genetic variance is high in ________ **Self-pollinated**________ crops.
10. Vertical resistance shows ________ **Oligogenic** ________ type of gene action.
11. The variance which is due to deviation from additive gene action resulting from intra-allelic interaction is ______ **Dominance variance**_______ .
12. For transfer of resistance gene governed by polygenes ________ **pedigree**________ method is used.
13. Pigeonpea is a ________ **Often cross**________ pollinated crop.
14. The 1st hybrid developed by C.T. Patel in cotton is ________ **H4** ________.
15. An example of protandry is ________ **Maize** ________ .
16. Source of male sterile gene in sorghum is ________ **kafir 60** ________.

17. Kind of rights are provided to authors for safekeeping their literatures is ________ **Copy right**.
18. ________ **Introduction** ________ is the quickest method of crop improvement.
19. Directorate of wheat research is situated at ___________ **Karnal** ________.
20. Bulk method was given by ________ **Nilsson-Ehle** ________ .
21. **Back cross method** is used for improvement of oligogenic traits.
22. The source of dwarfing gene in rice is **Dee-gee-Woo-Gen** ________ .
23. ________ **B-line** ________ is known as maintainer line.
24. ________ **Recipient** ________ parent is known as recurrent parent.
25. **Pedigree method** ________ provides breeder maximum opportunities to apply his skill and judgment during selection of plant.
26. Ear to row method of breeding was developed by ________ **Hopkins** ________ in maize.
27. CSSRI is situated at ________ **Karnal** ________ .
28. Colchicine is used for ________ **Chromosome doubling** ________ .
29. ________ **Additive** ________ variance is known as heritable fixable variance.
30. ________ **UV Rays** is an example of non-ionizing mutagenic radiation.
31. ________ **RAPD** ________ is an example of PCR based marker.
32. **Sugarbeet** is an example of autotriploid.
33. The total number of single cross hybrids can be developed where 6 inbred lines are used ________ **12** ________.
34. Full form of RAPD is-**Random Amplified Polymorphic DNA** ________.
35. Seeds stored in active collection for years ________ **10-15 years** ________.
36. Most commonly used culture medium in tissue culture is ________ **M.S medium** ________.
37. Production of microspore and megaspore is known as ____________ **Sporogenesis** ________.
38. Term ideotype was proposed by ________ **Donald** ________ in 1968.
39. Gene-for-gene hypothesis was given by-**Flor** ________.
40. A cross between F1(hybrid) and one of its parent is known as ________ **Back cross** ________.

41. The concept of embryo culture was given by ________ **Hanning** ________.
42. RMO-257 is a popular variety of ________ **Mothbean** ________ .
43. ________ **Synthetic varieties** ________ utilize only a part of heterosis.
44. Overdominance hypothesis was given by-**East & Shull (1908)** ________.
45. Heterosis calculated over better parent is called ________ **Heterobeltiosis** ________.
46. Pure line theory was proposed by ________ **Johannsen** ________.
47. Quality protein maize varieties essentially have ________ **Opaque-2** ________ gene.
48. Synthetic varieties are produced in ________ **Cross**-pollinated crops.
49. Homozygous line develops in cross pollinated crops by continuous selfing is known as ________ **Inbreds** ________ .
50. RFLP extends for–**Restriction Fragment Length Polymorphism** ________ .
51. ________ **Triploids** ________ are produced by crossing between tetra ploid and diploid.
52. Genetic male sterility is maintained by crossing————**Male sterile with heterozygous Male fertile** ________.
53. The term heterosis was coined by ________ **Shull** ________.
54. Bringing of wild species under human management is called— **Domestication** ________.
55. The term vertical and horizontal resistance was proposed by ________ **Vanderplank** ________.
56. The ratio of genotypic variance to the phenotypic variance is called ________ **Heritabilty** ________.
57. A group of plant originating by vegetative propagation from single plant is called ________ **Clones** ________.
58. ________ **Apomixis** ________ prevents segregating of genes in the subsequent generation of F_1 hybrids.
59. Double cross involves crossing of-**Two single crosses** ________.
60. Base analogues are ________ **Chemical Mutagens** ________.
61. A host pathogen reaction which leads to the death of infected cell is called-**Hypersensitivity.**

62. PGMS are being used for development of-**Rice** ________ hybrid.
63. Quantitative characters governed by ________ **Polygens** ________ .
64. In ________ **full sib family selection** ________ both the parents are common.
65. Nature of pollination in pigeonpea is-**Often cross pollination** ________.
66. Tift-23 A CGMS source of ________ **Pearlmillet** ________ .
67. Breeder seed is the progeny of ________ **Nucleus seed** ________ .
68. The colour of breeder seed tag is ________ **Golden yellow** ________.
69. ________ **Haberlandt** ________ is known as father of tissue culture.
70. ________ **Filter sterilization** ________ is used for sterilization of growth hormones.
71. Potato is ________ **Self pollinated** ________ crop.
72. Haploid plants may be obtained from ________ **Pollen culture** ________.
73. High inbreeding depression is observed in ________ **Cross pollinated crops**.
74. HHB-67 is a hybrid of ________ **Pearlmille**t ________.
75. The process of cleaning the explant is known as ________ **Surface sterilization** ________.
76. When an inbred is crossed with an open pollinated variety,it is known as-**Top cross** ________ .
77. Total sum of genes is called ________ **Germplasm** ________.
78. A trait which is controlled by polygene is called ________ **Quantitative character** ________.
79. When embryo develops from antipodal or synergid cells is called ________ **Apogamy** ________.
80. The pollen sterility which is controlled by nuclear gene is referred as-**Genetic male sterility**.
81. _____**Plant Introduction** _____ consists of cataloguing and evaluation.
82. In _____ **Hybrid varieties** _____ farmers have to purchase new seed every year.
83. Sudden heritable change in the base sequence is called **Mutation** _____.
84. ________ **CGMS** ________ type of male sterility is used in pearlmillet for the production of hybrid.

85. Cleistogamy promotes ________ **Self pollination** ________.

86. In the ________ **Back cross** ________ method,non current parent is used only single time.

87. ________ **Triticale** ________ is the first man made cereal.

88. The concept of centre of origin was proposed by _____ **N.I.Vavilov**_____.

89. Heterosis and inbreeding depression are present in ________ **Cross pollinated** ________ crops.

90. When a pyrimidine is replaced by purine and vice-versa, is known as ________ **Transversion** ________.

91. Heterosis is a synonym of ________ **Hybrid vigor** ________.

92. A cross between single cross and an inbred is called ________ **Three way cross** ________.

93. ________ **Totipotency** ________ is the capacity of a cell cultured *invitro* to regenerate in to a whole plant.

94. The term recurrent selection was proposed by ________ **Hull** ________.

95. ________ **Inbreeding**--- is the mating between closely related individuals.

96. Plant brought and grown to a new place,where it was not grown earlier, is known as ________ **Introduction** ________.

97. The lines which are identical in all respect,except for one gene,called ________ **Isogenic** ________.

98. The average performance of a genetic strain in a series of crosses,is known as ________ GCA ________.

99. The Asia minor centre of origin is also known as ________ **Pessian centre of origin** ________.

100. CRRI is located at **Cuttack** ________.

101. Land races are called ________ **Primitive cultivar** ________.

102. X- rays were first discovered by ________ **Roentgen** ________.

103. An individual with gametic chromosome is known as ________**Haploids** ________.

104. ________ **Recurrent parent** ________ is repeatedly used in back cross method.

105. In ________ **Overdominance** -heterosis,the heterozygous always superior than their homozygotes

106. ________ **UV-rays** ________ is an non-ionizing radiation.

107. ________ **Chasmogamy** ________ is the condition in which flowers open after fertilization.

108. Mating between two different species of the same or different genera of the same family is termed as ________ **Distant Hybridization** ________.

109. ________ **Pedigree method** ________ is suitable for breeding of horizontal resistance as back cross method.

110. ________ *Raphanobrassica* ________ was produced by crossing *B. oleracea* x *R. sativus.*

111. ________ **Apomixis** ________ is effective in fixing heterosis.

112. A ________ **Complete** ________ flower contains invariably all parts of flower.

113. ________ **Hybridization** ________ is the most commonly used method for creating variation.

114. ________ **Mass selection** ________ is most simple, quick and easy method of improving cross fertilizing crops.

115. ________ **Muton** ________ is the smallest of genetic material which mutates to produce a mutant phenotype.

116. In ________ **Sporophytic** ________ system of self-incompatibility in the pollen is determined by the genotype of the pollen producing parents.

117. NBPGR has ________ **10** ________ substations.

118. In ________ **Faculatative** ________ apomixis the ovule is capable of either sexual or apomictic reproduction.

119. Genetic male sterility is generally governed by ________ **Recessive gene** ________.

120. Cross pollination in pearlmillet occurs due to ________ **Protogyny** condition________.

121. Evaluation of worth of the plant on the basis of the performance of progeny is known as ________ **Progeny test-**.

122. Segregation starts from ________ $\mathbf{F_2}$ ________ generation.

123. Alpha particles consists of two ________ **Protons** ________ and two neutrons.

124. In mass selection plants are selected on the basis of ________ **Phenotype** ________.

125. ________ **Pusa jai kishan** ________ is a variety of mustard developed through somaclonal variations.

126. First copy right system was contributed by ________ **Johnnes Gutenberg** ________.

127. **Indian biological acts** ________, regulates assess of genetic resources.

128. Extended form of WIPO ________ **World Intellectual property organization** ________

129. The WTO is a successor of ________ **GATT** ________.

130. Literary work is protected under ________ **Copyright** ________.

131. PPVA and FR authority is located at ________ **IARI** ________ in New Delhi.

132. Marketing plan is ________ **Trade secretes** ________.

133. ________ **UPOV** ________ is the Internationl Organization for PVP.

134. ________ **Certification marks** ________ are used to define standards.

135. Madrid Agreement was concluded in ________ **1891** ________.

136. The registration for GI is valid for ________ **10 years** ________.

137. Trade mark registration office is situated at ________ **Chennai** ________.

138. International day for Biological Diversity is observed on________ **22**________ May.

139. Duration of registration for extant varieties is—**-15 years**—.

140. India becomes the member of the Paris Convention on ________ **Dec.,7,1998** ________.

141. ________ **Growth regulators** ________ are thermolabile chemicals.

142. ________ **Luxphotometer** ________ is used to measure the intensity of light.

143. Process of development of embryo from somatic cell is known as________ **Somatic embryogenesis** ________

144. Agar-agar is obtained from ________ **Sea weed** ________.

145. The full form of GMP is ________ **Good Manufacturing Practice** ________.

146. ________**Calcium alginate** ________ is used to make synthetic seed.

147. After 2-3 weeks,the medium culture are transferred to________ **Shoot multiplication medium** ________.

148. A somatic hybrid in which nucleus is derived from one parent and cytoplasm from both the parent is known as ________ **Cybrid** ________.

149. Regeneration of shoot from callus is known as ________ **Organogenesis** ________.

150. Bt gene contains ________ **crystal** ________protein.

151. Most commonly used carbon source in tissue culture medium is ——- **Sucrose** ________.

152. ________ **Embryo culture**________ is a useful tissue culture technique to attempt inter specific crosses.

153. The temperature of tissue culture is kept between **25±2°C** ________.

154. Haploid from pollen grains were first developed by ________ **Maheshwari and Guha** ________.

155. The most commonly preferred osmotic is ________ **Mannitol** ________.

156. The PCR technique was developed be ________**Kary Mullis** ________.

157. Culture of cell without cell wall is referred as ________**Protoplast culture** ________.

158. Plants containing foreign DNA are referred as ________ **Transgeneic plants** ________.

159. Plant part, which is used for regeneration in culture medium is known as ________ **Explant** ________.

160. The full form of PCR is ________ **Polymerase Chain Reaction** ________.

161. Bt gene is derived from ________ ***Bacillus thuringiensis*** ________.

162. NAA and IBA belongs to the ________ **Auxins** ________.

163. Liquid medium is essential for________**Suspension culture**________.

164. ________ **Cold chain** ________is necessary to transport *in vitro* plantlets.

165. Flavr-savr is a variety of ________ **Tomato** ________.

166. Generally media is infected by ________**Fungus** ________.

167. DNA molecule obtained from RNA is called ________**CDNA** ________.

168. ________ **Autoclave** ________is an instrument capable of sterilizing by steam under pressure.

169. The term apomixis was first used by ________ **Winkler** ________.

170. The reproduction occurs by apomictic means only is known as –**Obligate apomixis** ________.

171. The term apogamy was coined by ________ **Anton de Bary**________.

172. Diploid parthenogenesis has been reported in ________ ***Taraxacum*** ________.

173. The germplasm which is collected from other countries,is known as ________ **Exotic germplasm** ________.

174. Indirect selection for a desired plant phenotype on the basis of banding pattern of linked DNA markers is known as ________ **Marker Assisted Selection** ________.

175. Haploids are sterile due to lack of pairing partners of ________ **Chromosomes** ________.

176. The commercial banana is ________ **Autotriploid** ________.

177. RFLP markers are generated by ________**Restriction endonuclease** ________.

178. In maize, normal lines have been converted in to QPM through________ **MAS** ________.

179. Heterosis is partially exploited in ________ **Synthetic variety** ________.

180. The term genotypes and phenotypes were first used by________ **Johannsen (1911)**.

181. Quantitative characters are governed by ________**Poly genes** ________.

182. QTL stands for ________ **Quantitative Trait Loci** ________.

183. Single marker analysis is used for ________ **QTL** ________ mapping.

184. Golden rice is rich in ________ **Carotene** ________ and iron.

185. Inbreeding increases ________ **Homozygosity** ________.

186. In pedigree method individuals are selected from—-**F_2** and later generations.

187. The purpose of SSD method is to rapidly ________ **Advances** ________ generation.

188. Phenotypic assortative mating is useful in the isolation of ________ **Extreme phenotypes** ________.

189. Recurrent selection is effective to increase the frequency of desirable ________ **Alleles** ________ in the population.

190. Transgressive segregants are superior to ________ **Both** ________ the parents.

191. The purpose of alien addition generally is transfer of ________ **Disease resistance** ________.

192. Somatic mutation is also called ________ **Bud mutation** ________.

193. Clones can be maintained indefinitely through ________ **Asexual reproduction** ________.

194. Cytoplasmic male sterility in maize was discovered by ________ **Rhoades** ________

195. Single, self pollinated, homozygous plant is called as ________ **Pureline** ________.

196. ________ **Back cross method** ________ is used for transfer of oligogenic trait.

197. Polycross test is most useful in **Crosspollinated crops** ________.

198. The cross between inbred and open pollinated variety is called as ________ **Top cross** ________.

199. Adverse effect of host plant on development of insect pest which feed on plant is called as ________ **Antibiosis** ________

200. ________ **Kalyan Sona** ________ is an example of primary introduction.

201. Source of cytoplasmic male sterile gene in sorghum is ________ **Kafir-60** ________.

202. Ear to row method in maize is given ________ **Hopkins** ________ .

203. The concentration of ________ **Proline** ________ increases when the plants are exposed to drought condition.

204. Source of Dwarfing gene in rice was found in which country ________ **Taiwan** ________.

205. The difference between population mean and mean of selected plants is known as ________ **Genetic Advance** ________.

206. Which method is also known as evolutionary method of plant breeding ________ **Bulk method** ________.
207. Wheat variety MLKS 11 was developed by mixing seeds of ________ **8** ________ isogenic lines.
208. Grid Method was developed by ________ **Gardner** ________.
209. Heterobeltiosis is also known a ________ **Better parent heterosis** ________.
210. **Avoiding Contact** ________ is the most important stage for disease escape.
211. Polyploidy is induced through ________ **Colchicine** ________.
212. For corrections of defects in any popular variety ________ **Backcross method** ________ is used.
213. Bagging is done to ________ **Prevent contamination from foreign pollen** ________ .
214. ________ **Germplasm** ________is the sum total of all the genes.
215. Pusa 408 is a mutant variety of ________ **Gram** ________.
216. A synthetic variety can safely be grown for________ **4-5** ________ years.
217. Generally ________ **5-8** ________ good general combiner inbred lines are used to constitute the synthetic variety.
218. The composite variety of Brassica campestris is ________ **Composite-1** ________.
219. Simple recurrent does not measure the ________ **Combining ability** ________.
220. RSSCA was proposed by ________ **Hull** ________ in 1945.
221. Denaturation process is done at ________ **94°C** ________for one minute.
222. Annealing is done at ________ **53°C** for 30 seconds
223. AFLP is based on ________ **Restriction enzyme & PCR** ________.
224. The term biotechnology was coined by ________ **Karl Erkey** ________ in 1919.
225. ________ **MAS** ________ is a indirect selection process.
226. ________ **QTL** ________ is a gene region that affects a quantitative trait.

227. **Chimera** ________ are the plants consisting of two or more genetically different somatic tissues.

228. An ________ **electron** ________ is an atom carrying only negative charge.

229. ________ **Recurrent irradiation** ________ is the repeated irradiation of biological material in one or more generation.

230. ________ **Temperature sensitive mutants** ________ behave like wild type cells at low temperature.

231. Addition and deletion together leads to **Frame shift mutation**________ ______.

232. ________ **Temperature sensitive mutants** ________ are able to survive at specific temperature range not on other.

233. The exchange of purine by pyrimidine or pyrimidine by purine is known as ________ **Transversion.**

234. **Ancon breed of sheep** ________ is a classical example of germinal mutations.

235. **Swaminathan** ________ (1965) indentified macromutations.

236. **Bacquerel** ________ (1996) invented radioactivity.

237. ________ **Carriere** ________ (1865) referred to the sudden somatic changes as accidents.

238. The mutations producing drastic morphological changes in the phenotypes are called ________ **Macromutation** ________.

239. **Sub-vital mutations** ________ reduce the chances of survival by less than 50%.

240. **Missense mutation**-codes for different amino acids.

241. L1 gives rise to ________ **Epidermis** ________.

242. IIT-48 is mutant variety of ________ **rice** ________.

243. After tautomeric shift to its enol form, 5Bu pairs with ________**Guanine** ________.

244. ________ **Nitrosoguaninedine**-produces clusters of closely linked mutation.

C. Short Question and Answer

Q. 1. What is plant breeding.

Ans. Plant breeding is the art, science and technology of improving genetic make-up of crop plants in relation to their economic use for mankind.

Q. 2. Enlist different objectives of plant breeding.

Ans.
i) Higher seed yield
ii) Improved quality parameters
iii) Resistance to biotic stresses
iv) Resistance to abiotic stresses
v) Earliness
vi) Synchronous maturity
vii) Photo insensitivity
viii) Profusing tillering

Q. 3. What are important areas of plant breeding.

Ans. There are three main areas of plant breeding as given below:
i) Plant genetic resources or germplasm
ii) Plant breeding techniques
iii) Seed production techniques.

Q. 4. Define composite variety.

Ans. In cross pollinated species, composite variety is a mixture of several genotypes (in equal quantity) which are similar in flowering, maturity, height, seed size, colour etc. is called composite variety.

Q. 5. Define synthetic variety.

Ans. A synthetic variety is developed by crossing in all possible combinations a number of inbred lines with good general combining ability and mixing their seeds (seeds of F_1's) in equal quantity is referred to as synthetic variety. It is relevant to cross pollinated species.

Q. 6. What are main features of composites.

Ans. Main features of composites are given below:
i) Component genotypes composites are open pollinated.
ii) They are constitute only to cross pollinated species.
iii) They are maintained by open pollination.
iv) The exploit heterosis partially.

v) Exact reconstitution is not possible.

vi) Highly adaptable to environmental changes.

Q. 7. What is germplasm.

Ans. Sum total of all the genes & their alleles in a plant is called germplasm. Germplasm is a collection of individual plants that contains a species' genetic variation.

Q. 8. What is phenotype?

Ans. The observable physical or biochemical characteristics of an organism is known as phenotype.

Q. 9. Explain QTL.

Ans. *Quantitative Trait Locus*. This is a location of a chromosome region which contributes significantly to the overall phenotypes of a quantitative trait. It may be (1) a locus that influences the expression of a quantitative trait, or (2) a chromosome region detected by statistical analysis that is significantly associated with variation for a quantitative trait.

Q. 10. Why heterosis is partially exploited in synthetics variety?

Ans. In synthetic varieties, heterosis is partially exploited because some level of inbreeding takes place due to open pollination in later generations.

Q. 11. What is yield level of synthetic variety?

Ans. The yield of synthetic variety is always higher than open pollinated variety but lower than the yield of single and double cross hybrids.

Q. 12. What is transgenic breeding?

Ans. A branch of plant breeding which deals with genetic improvement of crop plants for various economic characters through plant biotechnology (genetic engineering) is known as transgenic breeding.

Q. 13. What is molecular breeding?

Ans. A branch of plant breeding which deals with genetic improvement of crop plants for various economic characters through indirect selection through linked molecular markers (DNA markers) is called molecular breeding. It utilizes molecular techniques such as RFLP, AFLP, RAPD, CAPS, and SSR etc. for selection of superior plants.

Q. 14. What is maintenance breeding.

Ans. It deals with ways and means of maintaining genetic and physical purity of released and notified varieties and parents of commercial hybrids.

Q. 15. Why yield of synthetics is poorer than hybrids.

Ans. The yield of synthetics is poorer than hybrids, because in synthetics heterosis is partially exploited, whereas in hybrids heterosis is fully exploited.

Q. 16. Why synthetics have better resistance to diseases.

Ans. Synthetic varieties have better resistance to plant diseases due to their heterogeneous nature and broad genetic base.

Q. 17. Is it possible to reconstitute a composite variety.

Ans. It is not possible to reconstitute a composite variety, because such variety is developed from heterozygous genotypes

Q. 18. What are the benefits of Biotechnology?

Ans. This technology helps to develop plants for pest resistance, to tolerate the stressful conditions like low temperature, drought, salt in soil, etc., helps to generate vaccines for the animals to fight against diseases, it is also used to produce pharmaceuticals etc.

Q. 19. What is Electrophoresis.

Ans. Electrophoresis is also known as Cataphoresis which shows the existence of an electric charge. This method is used in molecular biology and medicine and can be defined as the movement of colloidal particles under the influence of an electric field. When colloidal particles reach the oppositely charged electrode they get neutralized and coagulated.

Q. 20. What is Tissue culture.

Ans. Tissue Culture is an artificial growth of tissues or cells in a medium which is derived from living tissue or an organism.

Q. 21. What is 'Restriction Enzyme.

Ans. Restriction endonucleases are bacterial enzymes that bind and cleave DNA at specific target sequences. Type II restriction endonucleases are the most widely used in molecular biology applications. These enzymes bind DNA at specific recognition site, consisting of a short palindromic sequence, and cleave within this site. Restriction endonucleases with shorter recognition sequences cut more frequently than those with larger recognition sequences.

Q. 22. What is the role of Agarose Gel Electrophoresis in DNA.

Ans. Agarose gel electrophoresis allows separation, identification and purification of nucleic acids based on charge migration. Migration of nucleic acid molecules in an electric field is determined by size and conformation, allowing separation of different size nucleic acid fragments. However, the relationship between fragment size and rate of migration is non-linear, since larger DNA fragments have greater frictional drag and are less efficient at migrating through the polymer.

Q. 23. What is DNA?

Ans. DNA is the genetic material except for some viruses where RNA is the genetic material. DNA and RNA belong to a category of macromolecules known as nucleic acids. The structure of DNA was proposed by Watson and Crick in the year 1953. A DNA molecule is a long chain polymer composed of monomeric units called nucleotides. Therefore, a DNA molecule is made of a polynucleotide chain. Each nucleotide is made of a deoxyribose sugar molecule, a phosphoric acid group and nitrogen bases.

Q. 24. What is the role of biotechnology in crop improvement?

Ans. Biotechnology plays important role in following ways:

i) In developing transgenic varieties with herbicide resistance, good quality and resistance to biotic and abiotic stresses.

ii) In marker aided selection.

iii) In overcoming barriers of cross incompatibility through somatic hybridization.

iv) In developing transgenic male sterility for use in hybrid seed production of different crops.

Q. 25. What are the uses of apomixis in plant breeding.

Ans.

1. Rapid multiplication of genetically uniform individuals can be achieved without risk of segregation.
2. Heterosis or hybrid vigour can permanently be fixed in crop plants, thus no problem for recurring seed production of F_1 hybrids.
3. Efficient exploitation of maternal effect, if present, is possible from generation to generation.
4. Homozygous inbred lines, as in corn, can be rapidly developed as they produce sectors of diploid tissues and occasional fertile gametes and seeds.

Q. 26. Define inbreeding, inbreeding coefficient and inbreeding depression.

Ans. It is the mating between individuals related by descent or having common ancestry. The highest degree of inbreding is obtained by selfing. The degree of inbreeding is measured by the inbreeding coefficient. Inbreeding depression is the reduction in yield and viogour due to contineous selfing.

Q. 27. Define allopolyploidy.

Ans. Polyploid having chromosome sets from different sources such as different species; a polyploid that contains genetically different chromosome sets e.g., wheat, cotton, oilseedrape, peanut.

Q. 28. What is single-seed descent (SSD).

Ans. Deriving a population of inbred plants by self-pollinating individuals for several generations. Each individual from a segregating population, such as an F_2, is self-pollinated and seed are collected separately for each individual. A single seed from each individual is sown, self-pollinated, and seed are collected separately in each generation. After 5-6 generations each line is considered inbred.

Q. 29. What is typical flower

Ans. The reproductive structure in flowers. The flowers commonly consists of four organs namely sepsis, petals, stamens (male organs) and pistil (female organ). Both stamens and pistil are known as reproductive parts.

Q. 30. What is introgressive breeding?

Ans. It deals with genetic improvement of crop plants by transferring desirable genes from wild species into cultivated species of crop plants.

Q. 31. Who identified self and cross pollinated plants?

Ans. In plants, self-pollination and cross pollination were first identified by Sprengel in 18th century.

Q. 32. Who developed semi-dwarf varieties of wheat?

Ans. Semi dwarf varieties of wheat were first developed by Norman E. Borlaug in 1965. He developed semi-dwarf varieties of wheat at CIMMYT in Mexico.

D. Write Short Notes

1. Additive genetic variance
2. IPR
3. Wheat ideotype
4. Drought Hardening

5. Classify and explain types of seed collection on the basis of use and duration of conservation.
6. Explain ideotype of Wheat.
7. Define multiline and briefly explain its utility towards durable resistance .
8. Draw Schematic flow chart for pureline selection.
9. Enlist the features for developing Recurrent Selection for GCA.
10. Different mechanisms for developing insect resistance.
11. Differentiation between drought avoidance and drought tolerance.
12. Explain regarding Modified Ear to Row method in Maize.
13. List out different types of maize along with their scientific name.
14. Explain different measures of conservation of heterosis.
15. Differentiate between synthetics and composites.
16. Discuss the role of useful genes in crop improvement.
17. Briefly describe the achievements of clonal selection.
18. Briefly describe inbreeding and its effects.
19. Significanes of apomixis in plant breeding.
20. Discuss different classes of seed.
21. Explain in brief single seed descent method of selection.
22. Applications of autopolyploidy in crop improvement.
23. Explain reciprocal recurrent selection.
24. Explain in brief development of plant through meristem culture.
25. Briefly discuss the concept of centre of origin.
26. Explain dominance hypothesis of heterosis.
27. Differentiate between clone, pure line and inbreds.
28. Methods of over coming self incomptibility in crop plants.
29. Differentiate between heterosis and inbreeding depression.
30. Explain bulk method of breeding.
31. How mutation is useful in plant breeding.
32. Give evolution of wheat.
33. Use of PGMS and TGMS in heterosis breeding.
34. Write short note on half sib and full sib families.
35. How will you develop a plant through tissue culture.

36. Differentiate between polygens and oligogenes.
37. Briefly describe the techniques of anther culture.
38. Explain pure line theory of Johannsen.
39. Write short note on development and evaluation of inbred lines.
40. Elaborate the contribution of plant introduction in Indian Agriculture.
41. Discuss the merits and demerits of plant introduction.
42. Explain in brief recurrent selection for SCA.
43. Explain in brief mechanism of disease resistance.
44. Differentiate between primary and secondary centre of origin.
45. Differentiate between bulk and pedigree method.
46. What is simple recurrent selection.
47. Discuss gene for gene relationship.
48. Write short note in DUS.
49. What is Paris Convention.
50. Differentiate between vertical and horizontal resistance.

E. Long Questions

1. Explain about different mechanisms of disease resistance.
2. Explain genetic basis of heterosis.
3. Explain Hardy-Weinberg Law and factors affecting it.
4. Define Plant Breeding. Briefly describe various objectives of plant breeding with examples.
5. Outline the procedure for transfer of a dominant resistance gene through Backcross method. Mention the merits of this method.
6. Outline the procedure for transfer of a recessive resistance gene through Backcross method. Mention the demerits of this method.
7. Explain different activities in Plant Breeding.
8. What do you under stand by mutation? Briefly describe the procedure of mutation breeding .
9. Explain about different mechanisms of drought resistance.
10. Write down steps for identifying suitable clone and hybridization of the same.
11. Classify mutation on different basis of classification.
12. What is mutagen and Classify elaborately about physical and chemical mutagens.

13. Explain regarding polyploidy with suitable examples.
14. Draw a schematic diagram of pedigree method. Briefly explain its merits and demerits.
15. Explain in details about the steps involved in RAPD.
16. Briefly discuss the achievements made through mutation breeding in crop improvement.
17. What do you understand about alloploploidy? Explain their utilization in plant breeding.
18. Define self incompatibility? Discuss different types of SI with the help of suitable examples.
19. What is recurrent selection? Describe the recurrent selection for general combining ability in cross pollinated crops.
20. What is variety? Describe various operations involved in the release of a new variety.
21. Discuss maintenance of inbred lines, single hybrid production and commercial hybrid seed production for hybrid seed production in maize.
22. What is genetic engineering? What are the benefits of genetic engineering in crop plants?
23. Discuss heterosis and give various theories of heterosis and discuss its use in plant breeding.
24. What are synthetic and composite varieties? Discuss various steps involved in the development of synthetic varieties.
24. What is male sterility? Describe different types of male sterility.
25. What is ideotype breeding? Discuss genetic manipulation through recombining breeding.
26. Discuss selection method in clonally propagated crops with assumption and realities.
27. What are different methods used for improvement of self pollinated crops? Describe pedigree method of breeding along with merits and demetits.
28. What is hybrid ? Discuss the operations used for hybrid seed production of pearlmillet.
29. What is recurrent selection? Describe the procedure for improvement of two different populations simultaneously.
30. How the crops are classified on the basis of pollination behavior? Give example in each case. Also describe mechanism to promote self and cross pollinated crops.

31. What are reproduction and its different forms? What are merits and demerits of different kind of reproductionin the background of plant breeding?
32. What are autoploy ploidy? Describe the importance of polyplody in improvements of crop plants,giving suitable examples.
33. What do you understand by plant breeding? Discuss the major objectives of plant breeding.
34. Define transgenic? How gene transfer is done through particle gun method.
35. Describe the method of protoplast isolation, fusion and recovery of somatic hybrids.
36. What is apomixis? Discuss its type with the help of suitable examples and also focus on its significances.
37. What is alloploy ploidy? Briefly discuss the evolution of brassica spp. and wheat.
38. What is the meaning of Intellectual Property Rights? Briefly describe various types of IPR and legislations covering IPR in India.
39. What is meaning of patent? Describe about Indian patent system.
40. Describe the risks of commercial plant tissue culture and how to over come them.How to ensure that the plats produced from tissue culture are free from other pathogens.
41. What is nutrient media? Describe the major components of culture medium required for microproagation.
42. Define micropropagation? Describe the major steps along with its advantages and disadvantages of micropropagation.
43. What is sterilization? Discuss different sterilization techniques.
44. Define regeneration? Discuss the development of somatic embryo from single superficial cell of the explant.
45. What are stages of somatic embryogenesis? How can it be automated.
46. What are molecular markers? How MAS has been better than conventional breeding methods.
47. Highlights the importance of biotechnology in crop improvement.
48. What is PCR? Describe all the steps of this technique and its application in biotechnology.

46. How Bt cotton has been developed? Discuss all the necessary steps. Define vector? Briefly describe various kind of vector for *E. coli*..

47. Describe the recombinant DNA technology.
48. What is tissue culture? Discuss various steps used in tissue culture techniques.
49. Discuss safety rules and regulations in TCL.

Further Readings

1. Principles of Plant Genetics and Breeding – *George Acquaah*
2. Principles of Plant Breeding – *Robert W. Allard*
3. Principles and Procedures of Plant Breeding – *G. S. Chahal and S. S. Gosal*
4. Key Notes on Genetics and Plant Breeding – *Venkata R. Prakash Reddy*
5. Marker-Assisted Plant Breeding: Principles and Practices – *B.D. Singh, A.K. Singh.*
6. Plant Breeding and Cultivar Development – *D.P. Singh, A.K. Singh, A. Singh*
7. Plant Breeding and Genetics – *Hari Har Ram*
8. Plant Breeding: Principles and Prospects- *M.D. Hayward, N.O. Bosemark, T. Romagosa*
9. Principles of Cultivar Development: Theory and Technique – *Walter R. Feh·*
10. Plant Mutation Breeding and Biotechnology- *Q. Y. Shu, Brian P. Forster, H. Nakagawa*
11. Horticultural Plant Breeding – *Thomas J. Orton*
12. Plant Breeding for the Home Gardener: How to Create Unique – *Joseph Tychonievich.*
13. Breeding Field Crops – *John M. Poehlman*
14. Plant Breeding – *Jack Brown, Peter Caligari, Hugo Campos*
15. Statistical and Biometrical Techniques in Plant Breeding – *Jawahar R. Sharma*
16. Hybrid: The History and Science of Plant Breeding – *Noel Kingsbury*
17. Plant breeding: Classical to Modern – *P.M. Priyadarshan*
18. Dictionary of Plant Breeding – *Rolf H. J. Schlegel*
19. Quantitative Genetics, Genomics and Plant Breeding, 2nd Edition – *Manjit S. Kang.*
20. Next Generation Plant Breeding – *Yelda Ozden Çiftçi*
21. Advances in Plant Breeding Strategies: Breeding – *Jameel M. Al-Khayri, Shri Mohan Jain, Dennis V. Johnson*
22. Principles of Plant Genetics and Breeding – *Nina Duran*
23. Rediscovery of Landraces as a Resource for the Future – *Oscar Grillo*
24. In Vitro Plant Breeding – *Acram Taji, Prakash Kumar, Prakash Lakshmanan.*
25. Plant breeding methods – *Maha Bal Ram.*
26. An Introduction to Plant Breeding – *Jack Brown, Peter Caligari*
27. Quantitative Genetics and Selection in Plant Breeding – *Günter Wricke, W. Eberhard Weber·*
28. Selection Methods in Plant Breeding – *Izak Bos, Peter Caligari*
29. Accelerated Plant Breeding, Volume 1: Cereal Crops – *Satbir Singh Gosal, Shabir Hussain Wani*
30. Mutation Breeding: Theory and Practical Applications – *A.M. van Harten*
31. Experiments in Plant Hybridization – *Gregor Mendel*
32. Plant Breeding: Past, Present and Future – *John E. Bradshaw*
33. Organic Crop Breeding – *Edith T. Lammerts van Bueren, James R. Myers.*
34. Crop Improvement: New Approaches and Modern Techniques – *Khalid Rehman Hakeem, Parvaiz Ahmad, Munir Ozturk*
35. Plant Breeding and Genetics At A Glance – *I.D. Tyagi*
36. Plant Breeding, Theory and Practice – *V.L. Chopra*
37. Hybridization of Crop Plants – *Walter R. Fehr, Henry Hultman Hadley*

38. Molecular Plant Breeding – *Yunbi Xu*
39. Breeding for Quantitative Traits in Plants – *Rex Novero Bernardo*
40. Plant Molecular Breeding – *H. John Newbury*
41. Fundamentals of Plant Genetics and Breeding – *James R. Welsh*
42. Essentials of Plant Breeding – *Rex Bernardo*
43. Practical Plant Breeding – *S.K. Gupta*
44. Principles of Crop Improvement – *Simmonds*
45. Plant Breeding II – *Kenneth J. Frey*
46. Plant Breeding Principles and Methods – *B.D. Singh*
47. Fundamentals of Plant Breeding – *Phundan Singh*
48. Hand book of Genetics and Plant Breeding – *Rajendra Kumar Yadav*
49. Introductory Principles of Plant Breeding – *R.C. Choudhary*
50. Heterosis Breeding – *B.Rai*
51. Crop Improvement and Mutation Breeding – *A.K. Sharma*
52. Breeding Technology of Crop Plants – *A.K. Sharma*
53. Plant Breeding – *Sukumar Dana*
54. General Plant Breeding – *A.R. Dabholkar*
55. Hybrid Cutivar Development – *S.S. Banga and S.K.Banga*